N·O·R·T·H·W·E·S·T C·O·A·S·T

Vine maple and frost.

Soleduck River patterns, Washington.

The Olympic Mountains, near Mildred Lakes, Mount Skokomish Wilderness, Washington.

NORTHWEST COAST

Essays And Images From The Columbia River To The Cook Inlet

WRITTEN BY BRADFORD MATSEN, PHOTOGRAPHED BY PAT O'HARA,
PRODUCED BY MCQUISTON & PARTNERS

THUNDER BAY PRESS, SAN DIEGO

This book is about that territory of North America most blessed by the physical implications of water, a seventeen-hundred–mile stretch of the continent merging with the Pacific Ocean in a chorus of land fragments and fractures.

For Brad and Mae Matsen, my parents

Many people provided me with information, support, and hospitality during the writing of this book. I want to thank Laara Estelle, Krysten Holmes, Joel Bennett and Luisa Stoughton, Mark Brinster and Sue Hoffman, Frank and Donna Caldwell, David and Maureen Crosby, Patricia Guiget, Alan Haig–Brown, Bruce Hart, Holly Hughes, Margy Johnson, Bonnie Kaden, Hayden Kaden, Jean–Paul Kissel and La Rive Gauche, Pat Lane, Max and Heather McCarty, Barre McClelland and Sheila Murphy, Melanie Markle, Noel Pallas, Ron and Lorna Ross, Roald Simonson, John and Rosalie Simpson, Bill Spear and Susan Kirkness, Clem and Diana Tillion, Vincent Tillion and Tracey Harpole, Han Timmers, and Bob Thorstenson.

I am also indebted to my tolerant colleagues at *National Fisherman* magazine, Pat O'Hara, my publisher Thunder Bay Press, and in Del Mar, Don McQuiston, Joyce Sweet, and Robin Witkin.

— BRADFORD MATSEN

A very *special* thanks to the staff and guides at Alaska Discovery, Juneau, Alaska, for their spirited support of this book project.

A gesture of appreciation goes to the following organizations and individuals: Glacier Bay Sea Kayaks, Barlett Cove, Alaska; Glacier Bay Lodge, Barlett Cove, Alaska; Gustavus Inn, Gustavus, Alaska; Haines Air Service, Haines, Alaska; Wrangell Air Service, Wrangell, Alaska; Harbor Air Service, Seward, Alaska; Gulkana Air Service, Gulkana, Alaska; Janelle Eklund and Frank Bird, Copper Center, Alaska; and to Kieth Lazelle, Rhan Flatlin, and Kennan Harvey for their field assistance.

— PAT O'HARA

Printed in Hong Kong by Dai Nippon Printing Co., Ltd.

Published by Thunder Bay Press
5880 Oberlin Drive
San Diego, CA 92121

Library of Congress Cataloging-in-Publication Data:
Matsen, Bradford. 1944—
Northwest coast: essays and images from the Columbia River to the Cook Inlet/written by Bradford Matsen; photography by Pat O'Hara; produced by McQuiston & Partners.
p. cm
Includes bibliographical references.
1. Northwest Coast of North America — Description and travel.
2. Matsen, Bradford — Journeys — Northwest Coast of North America.
I. O'Hara, Pat, 1947 — II. McQuiston & Partners.
III. Title.
F852.3.M378 1991 917.9504'43 — dc20 91-7439 CIP
ISBN 0-934429-78-2
10 9 8 7 6 5 4 3 2 1

Photo Credits
Front Cover: Totem, Bight State Historic Park, Ketchikan, Alaska
Back Cover: Totem, Sitka National Historic Park, Alaska
Frontispiece: Sea lions, Vancouver Island, British Columbia
Page 9: Birch tree, spring design

Morning light on Mount Rainier, looking across the White River, Mount Rainier National Park, Washington.

Nisqually Glacier, in Mount Rainier National Park, Washington.

The view across Ingalls Lake of Mount Stuart in the Alpine Lakes Wilderness Area, Washington.

Snow covered Black cottonwoods, one of the primary streamside trees of the Cascade Range.

Sunset light, southeast coast of Alaska.

F I N D I N G   M Y   R A N G E

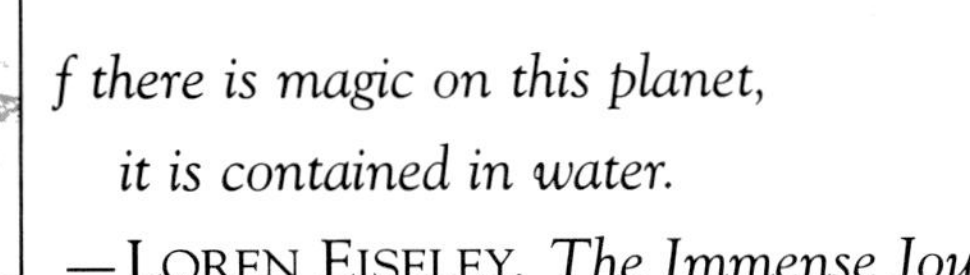

f there is magic on this planet,
it is contained in water.
—Loren Eiseley, *The Immense Journey*

Water revealed itself to me one August morning when I came on deck to finish my coffee and ease into the chores that begin every day of salmon fishing, a time that lends itself to reflection. I was at anchor in the lee of Point Adolphus, an idyllic promontory directly across Icy Strait from the mouth of Glacier Bay where the Pacific invades the northern reaches of Southeast Alaska through a gap just a few hundred yards wide and, at speeds of up to eight knots, encounters its first major terrestrial intrusion at Point Adolphus. When the tide peaks, especially at dawn as it did that day, the collision produces a violent churning of nutrients and a symphony of feeding fish, birds, and mammals. The fishing, therefore, was often great, especially when accompanied by the primitive pleasure of knowing that I was just another creature clinging to the very visible marine food web. I found enormous reassurance in nature's delivery of a message of abundance rather than travail.

The sensory snapshots that evoked the morning's thoughts of water carried the whistling sound of the ocean gathering itself into tide rips and maelstroms of astonishing power and the more modest splashes of herring, needlefish, and salmon breaking the surface in their frenzy. I can still see the synchronized flocks of northern phalaropes flashing alternately brown and white in the sky as though to the tapping toe of a divine choreographer who occasionally settles them as one into undulating rafts on the chrome black sea. Beyond the birds, a few miles across Icy Strait through clearing tendrils of fog, are the peaks of the Fairweather Range and the icy uplands of Glacier Bay, remnants of the sheet of frozen water that once covered not only the strait where I lay anchored but all the territory north to the pole and south to Puget Sound in Washington. And I recall, too, the fragrances of the damp forest washing across Point Adolphus to mingle with the pungency of life-created and life-gone that forms the tidal aroma.

I remember that morning like a card player remembers a pat hand because, on the grandest scale imaginable except for the view from space, the water that gives our planet life was on display in its three remarkably abundant states: liquid, gas, and solid. If some internal evolutionary force propelled me toward abundance for survival, that morning's vision of water validated the continental wanderings that had brought me to what has become my natural range between the waters of the Columbia River on the Oregon–Washington border and Cook Inlet in Alaska.

Overlooking the miraculous qualities of water is quite understandable because it is everywhere and a lot of people prefer sunlight — water's partner in earth's creative outburst — to rain. But with the exception of the air we breathe, water is the most important single substance to human beings, and its properties are amazing. Water is odorless, colorless, and tasteless, a compound of great stability, and a powerful solvent that is repelled by most organic substances but attracted by inorganic materials including itself.

Chief among the oddities arising from the consideration of water is its physical structure: the combination of two atoms of hydrogen and one of oxygen into a very sturdy molecule that, until a couple of hundred years ago, was considered an indivisible element rather than a blend of the stuff of stars and life. Nobody really knows exactly why water absorbs and releases heat more efficiently than almost any other naturally occurring molecule, and the temperatures at which water is transformed into its three states are not at all consistent with similar functions of other compounds. It expands when frozen, rather than contracting like most things, thus the solid is lighter than the liquid, which, of course, is why ice floats.

Water follows the general rules of heat transference in its liquid and gaseous states, but something happens when it approaches its freezing point. At 39F, water begins to get lighter, and when it freezes solid at 32, it has actually gained about 9 percent in volume, thereby reducing its weight. If water suddenly started behaving like just about everything else and weighed more frozen than as a liquid, life as we know it would depart because ice would sink. In the winter, more ice would accumulate and sink to the bottom of the oceans, and the summer sun would not be strong enough to melt the ice at such great depths. Life in the water would cease, and the seas would soon disappear. The hydrologic cycle on which our existence is based would grind to a halt.

Most remarkably, all life on earth depends for existence on the oddities of water to sustain the amazing harmony of the hydrologic cycle, a facet of nature clearly revealed along the Northwest Coast. In a complex but predictable pattern of climate owing to temperature, gravity, and the velocity of our rotating planet, water rises as a gas from the sea and land, condenses into liquid in the atmosphere, falls as rain and snow, and returns to the sea. The amount of water participating in the earth's hydrologic cycle has remained relatively constant since the last retreat of the continental glaciers.

About 99 percent of the water on earth is in the oceans and polar caps; the remainder is in rivers, glaciers, and groundwater, and at any given moment a small fraction is distributed through the atmosphere in the form of vapor or droplets. Rain. This amount is comparatively slight, and if all of it fell suddenly, it would cover the earth with only an inch more water. Once every twelve days, all the moisture in the air is recycled, and most of it returns directly to the sea.

Water is the paradise factor, held sacred in all cultures by all creatures. Twenty-five hundred years ago, Thales of Miletus first considered what he called the "genius" of water, claiming that the earth was alive by virtue of this one substance. In China, his contemporary Lao-tse came to the same conclusion; Taoism is indebted to the revelation that "Water is the Way," and for Buddha, the ocean was the clearest expression of Nirvana. Our connection to water extends beyond mere consideration of its nature because our body fluids contain precisely that proportion of salt present in the seawater at the supposed time our progenitors left it. And our food is not only mostly water but utterly dependent on it; a one-pound loaf of bread has used up about two tons of water while the wheat was growing.

You can step up to nearly anyone's front door the world over and get a free glass of water, but its inequitable distribution could well be the cause of the next great global conflict. Although the earth's biosphere seems to contain plenty of water to maintain the hydrologic cycle on average, some areas are better endowed with moisture than others, and in those places the most flourishing strains of life have evolved. A map showing the ocean basins, inlets, streams, and rivers from the Columbia River to Cook Inlet graphically explains why the Northwest Coast has nourished humans in a condition of superabundance for thousands of years, until recently when our presence has finally taken its toll on the region.

This book is about that territory of North America most blessed by the physical and spiritual implications of water, a seventeen-hundred-mile stretch of North American continent merging with the Pacific Ocean in a chorus of land fragments and fractures. It is as though the continent begins to come apart at the Columbia in the geologic rhumba of plate tectonics, becoming, with the exception of a few unbroken but still unstable stretches, a collection of islands and verdant watercourses that are attending servants to the big water of the ocean. The Northwest Coast is Waterland, and we have been granted a look at it in the twentieth century without the centuries of wear that have stripped Europe, eastern North America, and other densely populated regions of the world of much of the beauty and abundance that water begets.

The ice of the last glacial advance also unifies the Northwest Coast. The glacier's limits on the Pacific approximate those of this book, and the resurgence of complex life subsequent to its retreat create fairly even and typical flora and fauna throughout the region. The glacial scouring was a kind of pruning, it seems, and when the ice was gone, the hardy forests transformed the earth into a garden, with nitrogen-bearing plants and animals further assisting the evolutionary process of recovery and deterioration that continues even as you read this. The postglacial revival produced life in forms now considered typical of the Northwest Coast, such as the forests and the remarkable salmon, both of which receive considerable attention in the chapters that follow.

My natural range — the range of this book — is the northern extreme of the Pacific rain forest that we, in 1990, are just starting to recognize as an ecological treasure. We are children in this wilderness, only now maturing to our responsibilities to the environment, and the knowledge that allows us to do so has been slow in coming. Only during the 1980s did we come to understand that old-growth forest does far more than surrender top-grade wood products and offer the cosmetics of wilderness, although both of those contributions to our lives are valuable. The old forests purify water, store carbon, and, in a complex relationship with other life, stabilize the giant organism we call earth. The earth was once the forest planet as much as it is now the water planet, and we have no idea what we have done by clearing all the land for settlement and profit. Most of us, though, are starting to figure out that if we don't understand a natural system, the best thing we can do is leave it alone. The consequences of error literally could be the extinction of life.

The old-growth trees begin to dominate the landmass just north of San Francisco, though all but a ceremonial few have been logged as far north as the Columbia River, where we decided to begin this collection of essays and photographs. Some old-growth fir, spruce, and hemlock remain in Washington State, most notably on the Olympic Peninsula and in the North Cascades, and with the surge in public awareness of the value of this life-form to the earth's biosphere, these stands will remain exempt from extractive commerce. Less than 10 percent remains of the 20 million acres of old-growth forest that had covered California, Oregon, and Washington when the first European settlers arrived about 150 years ago.

Sadly, British Columbia's old growth is just about gone to market as well; the coastal parks of Vancouver Island and South Moresby in the Queen Charlottes are the only exceptions. In Alaska, most of the timber acreage is owned by the people of America, managed by the U.S. Forest Service, and finally under the kind of scrutiny it deserves. Already, substantial portions of the Tongass and Chugach national forests, which occupy all of Southeast Alaska and Prince William Sound, respectively, have been set aside.

The amazing salmon, another characteristic life-form whose natural range roughly defines the scope of this book, has also exerted a powerful influence on the natural and human history of the Northwest Coast. This postglacial descendant of the trout has the ability to leave the region, pack nourishment into its body, and deliver itself to feed just about every other form of plant

The Northwest coast is a waterland in a wide variety of forms. Vast snowfields, immense glaciers, and wild untamed rivers combine to make it one of the world's most scenic regions.

and animal in Waterland. California's salmon runs are recovering with help from hatcheries, and Oregon, with its precipitous coastline, depends on those artificial runs along with those of the Columbia River, whose banks Oregon shares with Washington.

The Columbia's salmon runs are coming back, with the help of hatcheries and conservative fishery management, but they will never reach the superabundance of the undammed river. In the 1930s, the construction of the Grand Coulee Dam to irrigate the Columbia Plateau and produce hydroelectric power cost billions of salmon, but the community of America made its choice and now we are living with it. The Columbia and its tributaries are now dammed in dozens of places; the more recent barriers provide for the fish with ladders and transport programs.

The Northwest Coast from the Columbia to Cook Inlet is the scene of a drama featuring nature in its pristine state and human beings propelled by the inevitable mechanics of population growth. At the end of the last global war, about 2 billion people lived on earth; in 1990, more than double that number shared the planet; in 2030, the number will double again to about 11 billion. There really is no question about what has happened and continues to happen to the Northwest Coast, one of the last zones of superabundance owing to sparse settlement.

When the Europeans arrived, however, they set to work with their enormous abilities and the strength of their Victorian vision that the earth and all its plants and animals existed to serve them. They further divided the ecosystem into political divisions that don't make much sense except in terms of commerce and the strains of nationalism that still remain even as we move toward a truly global village. Few people would disagree that from nature's standpoint, we would be better off with a single political entity—say, Waterland—instead of Washington, British Columbia, and Alaska.

Nonetheless, in part because of that sociopolitical history, this book on the Northwest Coast has been organized into three sections, and each section is the setting for a series of essays. The other reason for this organization has to do with my personal experience of twenty-five years of traveling in my range, which though it generally extends from the Columbia River to Cook Inlet, has as its center the wilderness of Southeast Alaska. The text and photographs are divided into three geographic sections: the Columbia River to Dixon Entrance at the British Columbia–Alaska border; Dixon Entrance to Cape Spencer, the southern and northern boundaries of Southeast Alaska; and Cape Spencer up the narrow strip of Alaska past Yakutat and across Prince William Sound to Cape Resurrection on the Kenai Peninsula.

This book is not intended to be didactic, although it is impossible to consider our history of interaction with this once-superabundant section of the coast without drawing some conclusions, as do the essays that follow. In a lecture given in Seattle in 1990, Brendan Gill of the *New Yorker*, a keen observer of human beings and our living habits, noted that we have a terrible history of moving into a beautiful area and making it unfit for life as quickly as possible. That some elemental curse or a role in the natural cycle that we do not fathom has fashioned us as the destroyers is not consistent with our creative individual sensibilities and does not accommodate the miracles we bring to life with our joy. There must be a way out of Gill's pessimism.

My purpose in these essays beyond giving you a sense of my range is to suggest that we pull back from further destruction of our habitat and find ways to live that don't depend on constant development and energy consumption for success. Shortly before his death, Albert Einstein was asked what he considered the most important question facing humanity, and he replied: "Is the universe friendly?" The remaining wilderness of the Northwest Coast nourishes life, elevates the human spirit beyond the concerns of commerce, and reassures us that the universe is indeed friendly.

ALASKA
Anchorage
Kenai Peninsula
Homer
COOK INLET
Kodiak
Kodiak Island
GULF OF ALASKA
Dawson
YUKON
NORTHWEST TERRITORIES
Haines Junction
Whitehorse
Haines
CAPE SPENCER
Juneau
BRITISH COLUMBIA
Fort Nelson
Sitka
PACIFIC OCEAN
Wrangell
Ketchikan
DIXON ENTRANCE
Prince Rupert
Queen Charlotte Islands
ALBERTA
Prince George
Bella Bella
Bella Coola
Vancouver Island
Vancouver
Victoria
Seattle
WASHINGTON
IDAHO
COLUMBIA RIVER
Portland
OREGON
THE NORTHWEST COAST

Mount Rainier, Washington, viewed from near Summit Lake.

COLUMBIA RIVER TO DIXON ENTRANCE

Evening light on Crown Point in the Columbia River Gorge National Scenic Area, Oregon.

R O O K I E S   O N   T H E   B A R

Picking our way across the Columbia River Bar, the lifeboat coxswain plays the flats and saddles between breaking fifteen-foot seas that rumble into this tentative edge of the continent. The waves are ragged brown humps sucking up acres of the bottom in each dose of fury, coating the surface with muddy foam, and suggesting a relentless disorder that will never subside. Much of the time, the coxswain's strategy of matching the rhythm of the waves fails, and he swings the bow into the advancing Pacific, crying a warning to the rest of us. We get ready, flexing our knees and leaning back in the harnesses that tie us to the boat, but when the wave hits, it always brings a startling shock and, for me, a moment of fright. Sometimes the wave that beats our plan is a rogue so vicious it is hard not to imagine some terrible purpose behind its progress. All the waves are rough business, though — a single cubic yard of water weighs almost a ton, and a fifteen-foot breaker will drop twelve hundred tons of water on our twenty-ton boat. Besides, most waves contain drifting logs and other debris.

Aboard the National Motor Lifeboat School's Forty-four Footer on a February afternoon, though, the point is to confront the terror of the most notorious river bar on the Pacific so that it will be no stranger during a real rescue. Three students — two from the Great Lakes and one from Monterey Bay — an instructor with a fifteen-year history of performing rescues at sea, and I are dressed in exposure suits, helmets, and the harnesses that lash us to the heaving boat. At this unique school, already competent mariners learn the skills for crossing deadly water to reach other mariners in distress and actually pluck a struggling swimmer from the surf or get a towline on a foundering boat.

The business of saving lives at sea has a venerable tradition, populated with heroes whose names are usually known only to the few whose lives have been handed back to them. Here, off Cape Disappointment, Washington, on the Columbia Bar, and at eighty-two other places in America that claim lives and boats with a deadly certainty, the Coast Guard stations its motor lifeboats. Here, too, they train the crews for those boats, because water just doesn't get much worse than on the Columbia Bar. Big surf is not benign, especially when encountered for the first time, but in an angry sea the power of the water is fully revealed.

Since 1961 the standard motor lifeboat has been this amazing

Forty-four, known just by her length designation, although each craft carries a name and registry number as well. These boats are identical: steel, self-bailing, self-righting, and virtually unsinkable owing to a series of watertight compartments and a low center of gravity. The Forty-fours are powered by twin diesel engines, with a modest top speed of about twelve knots, and although they're not fast, they can tow fairly large vessels through heavy seas. Of the 110 built, only one has ever been lost—off Auke Bay in Southeast Alaska when a vicious storm drove it up on the rocks—and it didn't actually sink.

The Forty-four looks a little like a floating pelican or other sea bird, a shape quite natural to its purpose. The Spartan steering cockpit amidships is the forehead, set behind a raised foredeck that could be the bill; the aft section bulges with an airtight "survivors' compartment," that, with a little imagination, becomes folded wings. The deck surfaces are convex to contribute to the boat's self-righting ability, an amazing characteristic of the Forty-four and its predecessors.

Though a capsizing is far from routine, it will not necessarily interrupt a rescue mission, or so says the manual for the motor lifeboat (MLB): "Properly balanced by outfitting and gear stowage, the MLB can heel a full 90 degrees and recover. If your lower gunwale digs into the trough, prepare to roll over. Take a deep breath and hang on." Right. "Any surf active enough to capsize the boat will bring it right back up. All that is needed is a 1.5 degree deflection to either side of the 180 degree position (i.e., rolled over and underwater) and the boat pops upright. The next wave will do it; average time will be eight to fifteen seconds underwater. There will be damage; the boat was built to be able to roll over, not to like it. If the engines are still working and there is no major hull damage, consider proceeding with the mission."

The hardware of modern surf rescues is really something, but I'm glad that I didn't read the manual before I was on the bar with the class of rookies. I guess eighteen seconds in the surf under a twenty-ton motor lifeboat is no problem for some people, but it is for me. That somebody does this for a living and is committed to routine occasions of such fear on behalf of their fellow human beings is inspiring.

Until the Forty-four came into service, the standard lifeboat was the Thirty-six, a single-screw, wooden lifeboat that was so reliable many felt its replacement would not measure up to the task. Before the Thirty-six, a variety of decked and canvas-covered displacement hulls under engine or oar power did the job from places whose names alone summon images of the wild sea: Cape Hatteras, North Carolina; Cape Cod, Massachusetts; Kodiak Island, Alaska. At the entrance to the lifeboat school grounds at Cape Disappointment, a Thirty-six rests on a cradle, anchor deployed and firmly fixed in the lawn, flag and ensign fluttering from the halyards. Beside the boat is a commemorative monolith, a wooden slab on which the following words are written in weathered, blistering gold paint:

On July 6, 1788, Capt. John Meares, a British fur trader, was proceeding along the Washington coast from Vancouver Island. He tried to cross the Columbia River Bar, and met with frustration and disappointment, hence he named the major headland here Cape Disappointment. On May 11, 1792, Capt. Robert Gray of the U.S. ship Columbia *made it over the bar and anchored off the site of what is now Chinook. The river is named for his ship. In 1850, Lt. William P. McArthur, U.S.N., surveyed the entrance to the Columbia River for a lighthouse site. Based on his recommendations, Congress approved $53,000 for the lighthouse on Cape Disappointment, completed in 1856. It is the oldest lighthouse on the Pacific Northwest Coast.*

In 1877, the Ft. Canby lifestation was established here under the Revenue Marine Service, the forerunner of the Coast Guard, and manned by local volunteers. In 1822, Capt. Al Harris swore in the first full-time crew. The Cape Disappointment Lifesaving Station has operated continuously since then during which time the Columbia River Bar has become known as "The Graveyard of the Pacific."

On a small boulder in the grass beside the historical marker is a bronze plaque with the names of the Coast Guard rescuers who died on the bar between 1961 and 1982—eight of them.

The passionate dissonance of the Columbia Bar belies the life-giving character of this miracle of a river. The Columbia River unifies the Pacific Northwest more than any other geographic feature, although the territory has now been subjected to political

divisions. The river gathers its waters from most of Idaho; nine-tenths of Oregon; more than half of Washington; parts of Montana, Utah, Wyoming, and Nevada; and an enormous section of southern British Columbia. The Columbia drains an area about the size of the state of Texas, and its flow exceeds the combined volume of all other rivers emptying into the Pacific between Canada and Mexico. It carries more than twice the water of the Missouri and ten times that of the Colorado.

And what is most consequential in the torrent of superlatives is the altitude drop between the Columbia's headwaters and the bar at Cape Disappointment. The river rises from the shoulders of the North American continent in the Columbia Lake bounded by the mile-high Selkirk Range of the Canadian Rockies and rushes along its main stream, fed by 150 tributaries big enough to be called rivers, to sea level about twelve hundred air miles to the west, covering twice that distance overland. Water that falls within a few miles of the Columbia's watershed also ends up in the Arctic and Atlantic oceans and the Gulf of Mexico through tributaries of other great drainages emanating from the same high terrain.

But few rivers on earth drop as far, as fast, and with as much volume as the Columbia, which carries more potential power than any other river except a few on the African continent. It far outstrips the Volga, another great power-carrier, and the Yukon, Ganges, Yangtze, and Amazon can't come close. The energy potential of the falling water of all the world's rivers has been estimated at about 750 million horsepower, 20 percent of which is in North America and a full third of that is in the Columbia.

The geologic history of the Columbia drainage, and therefore of the entire Pacific Northwest, is one of fire and ice, with a symphony of tectonics as eight separate plates, some of which have vanished almost entirely, crashed and withdrew over millions of years. The lithosphere — or crust — assembled by the moving plates was created almost entirely from portions of tilting oceanic trenches, with little actual continental crust.

The Columbia carved its first course through the basalt and lava of upheaval, once flowing more directly into the sea and, as recently as one hundred thousand years ago, carrying fifty times its present volume of water. Subsequent damming by emergent rock and, much later, by the waves of ice that flowed from the north forced the river into its present form, a looping meander that describes vast arcs. Once done with its initial wanderings at its headwaters, the Columbia follows the path of least resistance around a giant plateau composed of ancient sea bottom, which can easily be cleaved at its edges where harder rocks take over, until the coast. On the compass, the river flows northwest first, then hooks back to head generally south until midway into Washington where it takes a sharp turn to the west, past the great bend where we built the Grand Coulee Dam, then again heads south and even east, before swinging back west and through the Columbia River Gorge for its final run northwest to the Pacific.

The Columbia River Gorge tells the story of human interaction with the river and our ability to alter nature forever. This gorge, where the Columbia takes its final steep tumble to the sea through the hard young rocks of the coastal mountains, is probably the site of the longest continuous human settlement on earth. Archaeologists racing the builders of The Dalles Dam, which flooded that part of the river, found sixteen thousand years of human history revealed in potsherds, tools, and fishing implements. To the early people, the river's power was its ability to carry the salmon from the great pasture of the Pacific and to nourish the many other life-forms that could become food.

The electrical power drawn from the Columbia by the thirteen engineering marvels like The Dalles and Grand Coulee dams is another story. The epic hydroelectric and irrigation projects cost billions of dollars, billions of present and future salmon, many lives, and the original grandeur of the river's unfettered passage. But the community decision to spend so much money, food, and beauty essentially said that twentieth-century humans were willing to trade all of that for the river's power.

Using the Columbia to irrigate the central plateau also produced huge crops of wheat and corn where none had grown before. The builders were prodded by Manifest Destiny and the recent experience of the Industrial Revolution, which had taught them they could alter the earth on a grand scale to keep the engines of commerce running. The population nightmares of

Malthus and other visionaries were becoming galloping realities, and we were clumping ourselves together in great cities that could not be fed as the tribal enclaves had been. We were faced not only with producing more food, but with finding power of some kind to move that food far greater distances than ever before.

The Columbia is navigable by ocean-going ships up to The Dalles, just east of Portland, which brings the ocean inland a distance of 188 miles. From the first sightings by overland and seaborne white explorers, commercial traffic on the river was paramount in the minds of the newcomers, and they paid dearly to establish their visions in reality. The term *Columbia Bar* is a bit misleading, in fact, since it does not refer to a single predictable sandbar but to a collection of shifting, submerged, and emerging shoals of rock and silt washed from the river.

Unlike many rivers of such length and flow, the Columbia does not have a delta of much consequence like the Mississippi or the Yukon. In the flats around Longview, Washington, and west, the Columbia does wind a bit, and over the ages has dropped off oxbow lakes in its passage, but nothing like the distribution of material in a real delta occurs. So most of the solid matter the river carries ends up in the sea.

The bar, though, is not the main recipient of the Columbia's load. In the eddy off the river's mouth, the Pacific currents flow generally northward and, therefore, so does the river water and whatever it carries. Except for some hard-rock headlands like Cape Disappointment, the coast of Washington from Ilwaco seventy-five miles north to Grays Harbor is formed entirely from the outfall of the Columbia.

The natural patterns of the formative process between land, sea, and river were disturbed even here by the giant jetties built to keep a deep, navigable channel open across the bar. In the few years since the jetties went in, the formation of shoreline has accelerated just north of the bar on the Long Beach Peninsula. In some places, the accretion is so loose it is barely dry land at all and will likely turn to quicksand during the next large earthquake, after which the sea will again pound the sediment into the edge of the continent. All that might not be good for commerce, but so it goes.

On the bar in the Forty-four, we are using the Columbia's amazing hydraulics as a classroom. The instructor tells the rookies that the first rule of the rough-water coxswain is "don't let a wave break on you." Just about that time, one of the others in the crew shouts, "Starboard, over there. It's got a log in it," as a whistling fifteen-foot comber rises up, carrying some of the endless upriver debris on its flank. The man at the helm swings our bow into the advancing wave and hollers, "Hang on," as several thousand tons of the Columbia River and Pacific Ocean remind us who is really running things here.

COLUMBIA RIVER TO DIXON ENTRANCE

The power of the Columbia's rush from its mile–high beginnings to the Pacific is most vividly revealed in the Gorge, where steep terrain and numerous waterfalls guard the channel. The lodge and bridge at Multnomah Falls on the Oregon side were built in the 1920s, and for many, the memory of the 620-foot cataract — the fourth largest in America — forms an indelible image of wild hydraulics.

The flora of the Columbia Basin reflect the region's geologic youthfulness and recent scouring by ice, floods, and upheaval, all of which produce an exuberance in the variety of competitive forestation and ground cover. Spectacular seasonal displays of wildflowers and deciduous trees dance against the lush background of the evergreen rain forest, and over a thousand species of trees and plants make the area a botanist's dream.

The hot summers, moderate winters, and abundant water of the midriver region make the Columbia Basin perfect for fruit growing. This winter orchard on the White Salmon River at the east end of the Gorge, on the Washington side, shares earth and water with the hardy white oak, pine, and Douglas fir that lend their stabilizing influence to the valley slopes. The orchard country begins at about the White Salmon and runs along the western edge of the Columbia Plateau and into the North Cascades.

Human settlement in the Columbia Basin probably began 16,000 years ago. The Europeans started trickling in 150 years ago, bringing with them agricultural efficiency, along with their architecture. They also brought the fever of Manifest Destiny, a steady stream of migrants, industry, and an insatiable demand for hydropower, which led them to an orgy of dam building. Only a single fifty-mile stretch of the Columbia remains free-flowing in 1990, ironically along the banks occupied by the Hanford Nuclear Reservation.

Washington's Alpine Lakes Wilderness Area is a sliver of paradise between the North and South Cascades set aside in 1976 for "non-mechanized travel" by trail. Wildlife sightings and vistas such as this one of Prusik Peak in the Enchantment Lakes region are among the rewards for such voyagers. "Walking light" and "no-trace camping" are the rules to preserve the wilderness.

The splendor of a geologic statement like Dragontail Peak, near Aasgard Pass in the Alpine Lakes Wilderness Area, owes its existence to eons of terrestrial clashes during which eight separate pieces of the earth's continental and oceanic crust merged to form what we now call Washington State. The miraculous punctuations of flora including these blazing larch trees, or tamaracks, are as recent as a blink.

The Enchantment Lakes are among the most famous hiking destinations in Washington's North Cascades; the lakes are situated at seven thousand feet in granite basins surrounded by snowfields, glaciers, alpine meadows, and waterfalls. The individual lakes and tarns bear lyrical names like Rune, Talisman, Valkyrie, Naiad, Lorelei, Dryad, Pixie, and Gnome, bestowed by avid mountaineers and government topographers within the past hundred years.

Washington's Cascade River, here in its winter plumage, flows from the range crest into the Skagit River, the main watercourse of the North Cascades. The groundwater and icefield runoff from which these streams draw their flow also nourishes some of the earth's last remaining stands of old-growth fir, hemlock, and spruce.

The Wenatchee Mountains form the eastern slopes of the Alpine Lakes Wilderness Area, eventually slipping into the Columbia River where the mass of hard continental rock disappears beneath the vast inland plateau that occupies the western half of Washington. Swauk Creek, like hundreds of tributaries leading to the mother river, eventually delivers its water to the Pacific. Such creek beds on the eastern slope of the Cascades were scoured and reformed when the last great ice sheet blocked the Columbia at the Grand Coulee, triggering a deluge lasting thousands of years.

The Nooksack River Valley, in North Cascades National Park, near the Washington–Canadian border.

ON THE TRAIL

On the spring morning I drove to Sedro-Woolley, Washington, for a day's hike with Han Timmers, we spent the best part of our first hour together sitting in the sun considering his pet bees. Actually, consideration doesn't describe what we did, because he proposed a game of the senses involving his bees and the Bushmen of the Kalahari Desert who, by all accounts, possess the visual acumen to track a swarm of bees for a mile. "I have practiced with my bees," Han told me, "and I lose them in a hundred yards." I tried to follow the bees, too, as we sat there talking and picking through a bag of his bittersweet, home-dried apples, and I lost mine against the fair-weather blue sky just fifty feet or so from the trio of hives in the backyard.

Han's yard surrounds his two-story frame house, built at the turn of the century by a senior functionary of one of the logging companies that dominate the human history of the Skagit Valley towns along the western approach to the North Cascades. The croquet-grade green lawn is illuminated by a giant lavender coast rhododendron, the biggest I'd ever seen with a spread that at least equaled the distance I could spot a departing bee. (Incidentally, the coast rhododendron is the state flower of Washington.) The house is embraced by a stand of oaks, walnuts, chestnuts, and sycamores, one of which lends its lower limbs to Han's training ropes and ladders. In the sunny southwest corner is the garden he'd planted in the spring with his father, Jaap, who was visiting from Holland.

Along the garden perimeter, Han's cat, Tua, stalked an unseen prey, and languishing with us watching the bees was Lily, a dog with a sensate presence akin to that of lead dogs on sled teams. Lily is a lot like Han in that way dogs sometimes have of mirroring their humans. She is physically spindly but rawboned and quick, emotionally alert but calm with a generally kindly attitude toward whatever happens to be the business at hand. Once, in a backcountry snow camp high in the Cascades, a curious lynx drew within six feet of Han and Lily, and she just watched, conveying no fear or aggression. Lily has been Han's constant companion on trails, snowfields, and the rock of the North Cascades since he found her at the pound in the summer of 1986.

"It must be that the Bushmen have grown genetically superior with their eyes because it is so hard to find food in Botswana and

the bees can lead them, or something like that," Han speculated aloud, interrupting my sedentary tour of his yard. I'd given up on the bees well before Han and agreed that something attributable to evolution or cosmic purpose must have given the Bushmen their keen vision. But, I pointed out (dragging us back to why I had come to see him), a Bushman of the Kalahari probably couldn't climb to the 10,778-foot summit of Mount Baker and ski down in a single day. This is a routine feat for Han, but one he treats with profound gratitude because he lives in a place where such vertical wilderness is within easy reach.

Han Timmers began life in the steaming Indonesian archipelago with his colonialist parents, transported from their native Holland to the tin mines at Belitung during the waning years of the European empires. The family returned to Holland when Han was five, and by the time he entered adolescence, the mountains had caught his attention. "We took vacations in the Alps, hiking and staying in the huts," he said. "I can't put my finger on the exact time I was drawn to climbing, but I remember seeing some people on a face in the Dolomites when I was about ten, and they looked as if they were in a different world. I was fascinated but not really driven to climb until a few years later. In high school, I just happened to go to Belgium with the local alpine club, and found out that a steep rock brought out pure enthusiasm in me. It is something like being a child forever, I suppose."

We left Tua, the bees, and our interlude on the porch as I was beginning to unravel the details and motivations of this mountaineer's life, which has included long sojourns in the highest mountains on earth and an almost pole-to-pole exploration of the great cordillera that links the Americas from Tierra del Fuego at the tip of South America to Mount McKinley in Alaska. He has carried his life from the crowded Dutch lowlands to the Alps; Patagonia; Bolivia; the North American Rockies; the Fairweather, Wrangell, and Alaska ranges; The Himalaya, including two ascents of Mount Everest; and finally to a home in the valley that ties the Cascades to the Pacific Ocean.

So substantial is his climbing reputation that Fred Beckey, another European arrival and the acknowledged master mountaineer of the Cascades who has named many of the peaks, called Han when he heard the Dutchman was in the area. Han and Beckey are friends now, and the elder climber still calls from unusual places at odd hours. "The last time was two o'clock in the morning from a phone booth in Oklahoma, or some place like that," Han recalled.

Han was just over thirty when he arrived in the North Cascades in 1982. Earlier in our long-standing friendship, Han had told me that he had worried for much of his life about being one of the "Men Who Don't Fit In," which is the title of a Robert Service poem about hardy nomadic men who never find a place that they truly become part of. But like many émigrés to this territory, Han is an adventurer who finally found his place after wandering a world barely able to contain his craving for the revelation of nature in the extreme.

If you consider only the surface terrain, determining the boundaries of the paradisiacal region labeled the North Cascades can be confusing. The mountain subchain of which it is a part emerges on the western edge of the continent just north of where the major California fault zone bends into the sea at Cape Mendocino. The pressures and instability of the tectonic waltz produced the explosive beads of a volcanic necklace that begin with Mount Shasta and include Mounts Hood, Saint Helens, Rainier, and Baker. The naming of peaks and ranges on a grand scale is a recent phenomenon, and it was not until the Lewis and Clark expedition (1804–1806) that the Cascades were so dubbed, according to the expedition's records. The voyagers came upon the magnificent waterfall in the Great Gorge of the Columbia River (now inundated by the water behind the Bonneville Dam) and thereafter referred to the surrounding mountains as "the mountains by the Cascades" and eventually, "the Cascades."

Subsurface geologic mechanics, however, reveal the North Cascades not only as the geographic northern extreme of the Cascade Range, but as an ancient piece of a distinct subcontinent. The area we call the state of Washington was formed by the "docking" of eight individual pieces of the earth's moving crust, creating a mosaic of continental and oceanic rock that assumed its present form about 200 million years ago. Where these rafts of crust collided, a variety of features resulted as one raft slid beneath another or they butted end to end to create a volcanic fault zone like the Cascades.

A subsurface map of this intricate, gargantuan puzzle shows that the chain of fire, a zone of incredible collision and pressure between several sections of crust, was actually covered by the arriving North Cascade Subcontinent. The subcontinent became the surviving surface about forty miles north of Seattle; Mount Baker, a hundred miles north of Seattle, is actually an extruded remnant of the volcanic chain poking through the subcontinent's rock sheath. The geologic boundaries of the North Cascades also include the Okanogan Subcontinent remnant to the east, whose major feature, the Okanogan Trench, reveals the ancient collision, and the Columbia Plateau to the southeast, a massive feature composed of oceanic crust.

The Olympic Peninsula to the west of the North Cascade Subcontinent lays across a collapsed section of oceanic crust called the Puget Sound Lowland, and it is geologically and visually accurate to note the very different structure of the mountain guardians of the Puget Sound. The North Cascades are formed of wrinkled continental crust and the Olympics of oceanic crust extruded and turned to rubble; although both ranges celebrate the union between sky and land with soaring peaks, snow, and ice, they couldn't be more different.

On the ride with Han and Lily along the Skagit River into the North Cascades, I also saw the effects of repeated glaciation in the relatively near past. The most recent ice sheet came and went just ten thousand years ago, a heartbeat in geologic time and a critical influence on why humans revel in the fertility and abundance of the high-country rain forest. By far the dominant lifeforms in the resurgence of the nitrogen-bearing plants and animals following the glacial retreat are the trees of the evergreen forests — such as the Douglas fir, hemlock, and cedar — that now decorate the valley slopes to about four thousand feet. Above this point, soil creation and liquid water are too sparse to support the evergreens' existence.

The latest comers to the region, human beings, have seen fit to cut down about 80 percent of this forest, including great sections adjacent to the Skagit River. Sedro-Woolley, where Han Timmers now lives, took root as a logging center, and most of the early trail blazing at the turn of the century had to do with timber cutting, mining, or power generation at the two dams on the upper Skagit. Fortunately, we have withdrawn much of the North Cascades Subcontinent from extractive commerce, largely because the mountain wilderness presents hope and inspiration beyond the mere conduct of business.

Though human access to wilderness is a mixed blessing because it so often seems a prelude to destruction, Route 20 from tidewater through the North Cascades to the banks of the Columbia River is a magnificent road. The full passage was not accomplished until the 1970s, and even now the road is usually closed in the winter by snow. The route parallels the Skagit River to its highest elevations where it is dammed to form Ross and Diablo lakes, the earliest hydropower projects in the region. Han and I drove to Marblemount where the real altitude gain begins, and then turned off onto the Cascade River Road, which follows a tributary of the Skagit to a trailhead, in turn extending all the way to Lake Chelan on the eastern slopes. We munched more dried apples, and Han became increasingly animated as we proceeded into the heart of his range.

He told me he has advanced degrees in mathematics and psychology; in Holland, he worked in a think tank, studying the process of human perception of growth, before beginning his migration to the North Cascades. Han said he misses such work occasionally, but for the past ten years, his income has come from commercial fishing in Alaska, guiding trips in The Himalaya, the Fairweather Range, and the North Cascades, and, recently, appraising real estate in Sedro-Woolley.

His perspectives are distinctly global; he feels that as a species we are struggling with the shift from responsibility to nations to a sense of the earth as a whole. "When we talk about old-growth forest and spotted owls as indicators of the ecosystem's well-being, this is a global issue," he said, as we entered the rain forest on a narrowing gravel road. "It is not a matter of just affecting jobs in the Northwest, because we can no longer think in terms of ten more years of jobs in exchange for the old-growth forest that the world needs to exist. We have to find a way to help the people who are displaced as we try to meet global needs such as saving what's left of the forest, but we must save the earth first. If a global vote were taken on this issue, there is no question how the rest of the world would vote. I know for sure how the people I know in Europe would vote."

In his flight from an ecologically decimated and densely populated Europe, which is almost completely devoid of true wilderness, Han functioned as the human equivalent of a mine-shaft canary. If you want to know where the human spirit can flourish and soar, follow someone like Han Timmers. I followed him for a few hours of conversation and easy walking along the Cascade River that day, and over his shoulder, he told me about the contrast between his old home and his new.

"Climbing in the Alps was like climbing in an arena," he said of those years. "You can no longer really be alone in the mountains, and it's an entirely different approach to nature, more competitive and aggressive. The lack of wilderness changes the way people behave. Climbers do things like link ascents of very difficult terrain with flying down to the valleys with parasails and climbing the next mountain, then parasailing down, over and over, and then make it back to the bar in the evening to talk about it. At Chamonix [France] now, you see midair collisions between parasail climbers, and you hear the rescue choppers all the time. The place has become the death capital of the world in the summer."

Han maintained that parasailing has grown to cult proportions in Europe because just being in the mountains, given their worn condition, is not enough; people demand something like jumping off peaks with parasails or high-risk skiing. "They ski terrain over there that we once considered for technical climbing using ropes and fastenings. I tried to get parasailing classes going here, but people are just not as interested in something so risky to most of them."

The North Cascade trail rose and fell through the thermocline with its odd warm air up high and the chill of the water and shadows below, the old-growth forest muffling our footfalls and damping the rushing tremolo of the river. I paused to write in my notebook, we separated to enjoy the solitude, and then joined up with Lily keeping track of both of us as though in charge. Han told me a story about heading for Mount Shuksan, a neighbor of Baker, one morning and driving up through an active clearcut right as the crew was preparing to fell a six-hundred-year-old Douglas fir.

"I stopped and talked to them," he said. "Nice guys, and they told me what they were doing. I went on and skied up and down Shuksan, then in the evening drove back through the clearcut. The men were gone, but the tree was laying there in sections, with sap leaking from the ends of them. I stopped and stood a while, and it seemed as if the tree were weeping. I haven't been able to shake the sadness of that tree, six hundred years old and mutilated like that. In Switzerland they have every tree mapped because it is so valuable, standing to protect against avalanche and erosion. That Douglas fir would have been a national monument there.

"When people finally stand up to oppose something like destroying wilderness, it is not always a logical matter. It is mostly spiritual, from the heart and soul, not the mind. What better can we trust? I think everybody wants to live in harmony with their place when they get time to think about it, but nobody really knows why. I only worry for myself that I give enough back," Han mused as we paused in a patch of sun before heading back to town. He stood at complete ease, scratching Lily's head. "I live simply and try not to leave the waste from my life around, but I get so much pleasure it makes me wonder."

I suggested that somehow in the grand scheme of the cosmos the production of joy or bliss *is* the contribution life makes to the universe, that our main job is to preserve our own harmony and our habitat to continue to create such joy. Maybe, when we are in a condition of bliss, we emit a distinct particle or form of something like light that is as valuable as hydrogen or helium to the universe's existence, whatever its bounds or conditions. Maybe that's how we repay nature and the wilderness for our happiness.

We had a good laugh, and then just walked back to the road with Lily trotting between us.

For a century now, mountaineers and wanderers craving not only stupendous terrain but solitude have been drawn to Washington's Cascade and Olympic ranges from around the world. Whatcom Glacier on Mount Blum, in North Cascades National Park, reveals the result of the pressures and instabilities of the tectonic waltz, the recent ice age, and the ongoing geologic forge of volcanism that gives the region its wild, even threatening character.

The jewel in the volcanic necklace of the Cascade Range is 14,410-foot Mount Rainier, also the centerpiece of Mount Rainier National Park which was set aside in 1899 as the country's fifth such treasure. The looming giant was known as *Tahoma*—"The Mountain"—before the European settlers assumed dominance in the region of fire and ice. According to some variations of the creation myth of the early people, the female Tahoma migrated from the Olympic Peninsula because the mountains there were growing too fast and becoming too crowded.

Falls Creek and a splash of vine maple on the flank of Washington's Mount Rainier—an example of the lush, diverse fauna that occupy the great, slumbering volcano's slopes. The heavy rain, snowfall, and glaciers on Rainier support the luxuriant fauna as well as more than fifty varieties of mammals and 150 species of birds. Just sixty miles southeast of Seattle, the park's four hundred square miles offer hiking, climbing, skiing, and camping year-round. Rainier's last significant eruption occurred between 1820 and 1854.

The high-country of Olympic National Park is a large rugged wilderness area dominated by Mt. Olympus. The lower slopes are heavily forested with high meadows that in summer are a tapestry of wildflowers including avalanche lilies, glacier lilies, and lupine. There are over 60 glaciers within the park. Access to the high-country is by way of Hurricane Ridge or Deer Park, both of which offer motorists spectacular views.

The Pacific coast of the Olympic Peninsula is a symphony of pristine beaches, thundering water, and the wild shapes of headlands of soft ocean rock sculpted by fierce erosion. Most of the peninsula, including the entire seacoast, the lower rain forest slopes, and the inland heights are contained in the Olympic National Park. Substantial portions are reserved as sanctuaries for the aboriginal settlers, including the Ozette, Quillayute, Hoh, and Quinalt bands.

About 140 inches of rain a year and the moderate maritime climate combine to create temperate rain forest, and nowhere more spectacularly than on the Olympic Peninsula. Giant spruce and red cedar dominate in the unlogged, old-growth zones of the forest, with western hemlock, Douglas fir, and big-leaf maples interspersed. Dozens of species of epiphytes — nonparasitic plants, like the spectacular hanging mosses, that grow on other plants — further give the rain forest its unique character. Nine aboriginal bands were once the exclusive occupants of the rain forest lowlands of the peninsula. Now a comfortable lodge that is a base for many travelers carries the name of one band — the Quinalt.

The 898,000-acre inland portion of the Olympic National Park was created in 1938, and the fifty-mile-long beach reserve added in 1953. The headquarters for the park is in Port Angeles, a fishing port, harbor of refuge, and massive log shipment terminal on the Strait of Juan de Fuca, the boundary waters between Washington and British Columbia.

Rivers that drain from the central hump of the Olympic Mountains like the Bogachiel, Quillayute, Hoh, and Soleduck — here shown delivering perfection to a drift boat of fishermen — have assumed legendary status among anglers around the country. Although sadly diminished by habitat destruction from logging, dams, and fishing pressure, five species of salmon, steelhead, and cutthroat trout return annually to spawn in these picturesque streams.

The Sauk River, a tributary of the Skagit over on the Washington mainland across Puget Sound from the Olympic Peninsula, is another river of wide repute among fishermen. Fishing in most areas is limited to fly only, catch and release, due to drastic reductions in stocks of salmon and steelhead over the past four decades. With careful management, though, the Sauk and many streams on the peninsula are showing signs of recovery.

Many of the fertile waterways that tumble from the North Cascades emerge from the region of Glacier Peak, a 10,541-foot giant from which no less than fifty glaciers are visible on a clear day. The Glacier Peak Wilderness Area lies to the south of the North Cascades National Park and occupies the least-developed area in Washington State.

Discovery Harbor, Campbell River, British Columbia.

ROWING THE POOL

nless you know the story, what you see and hear at dawn along a particular stretch of water just beyond the mouth of the Campbell River in midsummer is difficult to comprehend. The place is known to locals as the Tyee Pool, after the summer-run chinook salmon accorded the name *tyee* when they exceed thirty pounds. *Tyee* means "chief" in the Chinook Indian jargon that grew from trade with the white men.

In the dim morning light the scene at the pool is obscured by a haze hanging low over the water, but the white shapes of small, graceful rowing boats — perhaps two dozen of them — gradually emerge. They move on what appear to be random courses, coasting in one direction and laboring in the other. The oars groan in their locks, a sound audible from the shore, punctuated by the occasional splash of a rough stroke and the more insistent tones of voices.

A pair of human shapes rises from each boat. They are bundled in the wrinkled yellows and oranges of rain gear, one of them swaying in the oarsman's rhythm, the other swiveled to face aft. And from each stern projects the length of what can only be a fishing rod, describing a bobbing arc against the gray sky. Across the rips and eddies of swiftly moving current, the air carries the scratchy salutes of feeding birds and the odors of tidal debris, kelp, and eelgrass.

On this particular morning, we happen to be upwind from the pulp mill a mile north and so are spared the noxious fumes of cellulose in its induced state of decay. The Campbell River and Vancouver Island have hosted loggers longer than fishermen, and the belching pulp mills are familiar landmarks. The big mill at Campbell River was built in the 1950s. Like other natural resources, timber is a slave to consumers; the paper for a single Sunday edition of the *New York Times*, for instance, requires cutting about seventy acres of forest.

The rowing boats on the pool are identical, about fourteen feet long, shaped much like the elegant Whitehall skiffs with a broad beam, low gunwales, and a transom flared like the upper half of a flattened hourglass. In profile, the lapstrake planking repeats the curve of the gunwales, which rise dramatically fore and aft from their lowest point amidships. The lines call to mind Victorian maritime designs and, as much as anything, contribute to the sense that some ethereal ritual — some ceremony of the past — is under way at the Tyee Pool.

The town of Campbell River, midway up the Inside Passage, declares itself the Salmon Fishing Capital of the World, and tens of thousands of true believers converge annually for a five-month-long frenzy. From all over the world, they settle in from May to September, numbing the otherwise quiet coastal village with hordes of cars, campers, and boat trailers. A daily flotilla of as many as a thousand outboard skiffs, cabin cruisers, and substantial yachts surrounds the dominant headland, Cape Mudge, which marks the northern end of the Strait of Georgia.

The sport fleet works in clusters on stretches of well-proven water — like Green Can, Copper Bluffs, Frenchman's Pool, and Whiskey Point — presenting an unforgettable sight to passers-by aboard commercial fishing boats, ferries, and cruise liners. Few territorial secrets are kept, since every tidebook you pick up in the town's many sporting goods stores contains a complete list of good fishing spots and the proper tide on which to be at each. There is plenty of production but little elegance in this industrial version of sport fishing, and the modest pacings of the rowing skiffs a few miles away off the Campbell River itself are even more of a contradiction against such a background.

This sport-fishing extravaganza owes its existence to annual runs of chinook and coho salmon that are pinched into a narrow passage between Vancouver Island and the British Columbia mainland. Commercial fishermen, too, take the chinook and coho, along with sockeye, chum, and pink salmon. This hundred-mile stretch of the Inside Passage is so narrow in places that the tidal currents reach twelve knots or more and mere feet separate ships from the shallows and reefs.

Seymore Narrows, a gap just a few miles north of Campbell River and the narrowest point on the Inside Passage, claimed more than a hundred ships and 120 lives between 1875 and 1958, when the passage was cleared by blasting a massive subsurface mountain from its center. Engineers dug tunnels under the seabed up into the middle of the mountain, called Ripple Rock, and blasted it apart with thirteen hundred tons of TNT in the largest human-made, nonnuclear explosion in history until that time.

The salmon swarm between the Strait of Georgia and Queen Charlotte Strait on their way to spawn in the streams and rivers that form a capillary system carrying life and nourishment. The runs are shadows of their former selves, though, since heavy clearcut logging, other habitat destruction, and fishing pressure have taken their toll.

Vancouver Island is a massive landform, 285 miles long from Beechy Head on the Strait of Juan de Fuca at its southern extreme to Cape Scott on Queen Charlotte Sound to the north. Just north of Campbell River, the island assumes its greatest breadth of 80 miles, intruding on the glacially fragmented mainland coast across an archipelago of smaller islands and narrow passages. On the mainland side, the island coast is relatively moderate, uninterrupted save for a few short rivers, including the Campbell, which is now a long lake behind a dam just a mile from tidewater.

The island is divided axially by a central mountain range that attains its greatest elevation of 7,219 feet at the summit of Mount Golden Hinde. The western coast captures the contrasts of high peaks soaring from sea level in rugged splendor. Vancouver's west side is repeatedly cleaved by bays, sounds, and inlets reaching nearly to its centerline in valleys of sublime beauty. The magnificent trails of Pacific Rim National Park and perfect conditions for sea kayaking have brought the ocean coast great status among adventurers.

Vancouver Island, and all of British Columbia, is geologically young, unlike most of the rest of Canada, which is formed of the 2-billion-year-old Precambrian shield. British Columbia is a region of active mountain building, part of one of the earth's most spectacular physical features, the cordilleran chain of mountains that extends almost unbroken from Cape Horn at the tip of South America to Point Barrow, Alaska. The colliding and subsuming of the tectonic plates forming the Pacific Basin and the North and South American continents makes these mountains, and buries the old crust in the process.

About ten thousand years ago, the most recent ice age formed the apparent geography of Vancouver Island and the coastal mainland. In its wake followed the pioneer plants, the dense forests of spruce, hemlock, and other conifers, and the salmon, all dependent on the watersheds that are constantly replenished by heavy annual rainfall. The human presence began with early aboriginal settlement in small numbers that placed little strain on the resources. But then came the Europeans who built a world culture in part by extracting food from the sea, timber, and minerals on an industrial scale.

The Campbell River region is representative of the competition for natural resources that has been typical on Vancouver Island since the turn of the century. Fishing, timber, mining, and tourism coexist on uneasy terms, and many well-intentioned people have tried to fashion equitable compromises that conserve those resources that can be renewed. Among these advocates and peacemakers was a famous local treasure, the writer Roderick Haig-Brown, who lived in a house he named Above Tide on the south bank of the Campbell. The house is the first beyond the estuarine section of the lower river, beyond which the high tides do not deliver seawater. He and his wife, Ann, raised their four children there, one of whom, after a full career teaching in the interior of British Columbia, became a writer and editor.

Coincidentally, and not without a measure of irony, that child, Alan, is the executive editor of separate magazines for the commercial fishing, logging, and tourism industries. He is an engaging man who loves stories, especially about boats and fishing. Alan spent his summers away from teaching in the bush country on the decks of salmon seiners in his home waters, and he guided sportsmen in his youth. He is an intense and good-natured man who seems amused most of the time.

Alan still tells of a summer morning when his father was casting into the shimmering Campbell right in front of Above Tide and the pristine calm was shattered by chain saws roaring to life right on the opposite bank. (This was not his favorite fishing spot by any means, but it was handy.)

"My father just simmered," Alan said many years later. "They were subdividing and developing the other bank of the river, and it was obvious that everything would change very quickly. My mother, though, was not as calm. She's been fighting the developers ever since."

Roderick Haig-Brown's books—*Return to the River*, *River Never Sleeps*, and many others—all contain his tactful but opinionated views on topics ranging from wildlife to Canadian nationalism. He wrote formal prose that allowed many writers of his region and generation to reveal themselves comfortably to readers, and he is valued not only for his sensibilities but for his very human, vulnerable presence in print. For many years, he was the magistrate of Campbell River, the highest judicial officer in a remote coastal town, serving not only justice but also many of the emotional needs of his constituency.

In 1961, Roderick Haig-Brown was commissioned by the British Columbia Natural Resources Conference to produce a book eventually published as *The Living Land*. This landmark work is a narrative treatment of twelve years of effort by the academics, field scientists, and other contributors to the conference. Despite the more recent trends toward attempting responsible compromise in resource conflicts, *The Living Land* has become obscure.

"The book is intended to stimulate the thinking of British Columbians about the proper use of their rich inheritance of natural resources," say the conference members in the preface. "It is hoped, also, that the book will serve to advance the cause and philosophy of conservation among the peoples of the world."

"My father was very proud of *The Living Land*," Alan told me. "I'm not sure that it had the effect he hoped it would, though." In the text, which the elder Haig-Brown said contained a "strong predilection for sharp comment, broad generalizations, and contentious ideas," he inventoried fisheries, timber, mining, energy, and recreational resources.

"Immense natural resources—vast natural wealth—inexhaustible natural resources—these phrases and many others like them have been used about practically every geographic and political division in North America," Roderick Haig-Brown wrote in *The Living Land*. "They have long been the treasured toys of promoters and boomers and boosters, the happy playthings of politicians, the sad and doubtful comfort of struggling settlers and ordinary working people. Like other good phrases, excessively used and abused, they have become fogged in meaning, remote from reality, suggesting only some vague form of future wealth that may be realized if certain things happen; something that ought to be looked after and probably is not being; something that ought to benefit everyone and probably will benefit few."

"He was awfully opinionated, and understood our dependency on the earth." Alan said of his father. "I rowed as a guide on the Tyee Pool when I was a teenager, but my father wasn't too keen on it because he didn't like trolling. He thought you didn't interact with the environment enough, so you didn't get to know it. He wanted everybody to act responsibly. And I guess he just liked river fishing more anyway." Alan chuckled in remembrance. The distance of years from his father, who died in 1976, has given Alan not only the freedom to write himself, but an easier pace with which to appreciate him.

"I remember the first time I went down the river in a canoe as a kid," Alan said, "and it was a very big deal to go to salt water. When I was about fifteen, I started working on the docks at one of the lodges, and eventually got a few lessons from a master guide on how to row the pool. Guiding was kind of a mixed thing for me, and my father, too. He guided when he had to but didn't really like it. It was just something to do, and I liked boats and the water. Ned Painter was around, and I guess you could say he invented the Tyee Pool."

And Ned Painter built the first of the graceful rowboats that now carry his name, although the most recent versions are made of fiberglass instead of wood. The Tyee Club was founded in the 1930s as one of those exclusive groups favored by fishermen who want to memorialize their achievements on the water. Like most such groups, the Tyee Club is defined and governed by sporting rules that allow entry only to those who catch a particular fish in a particular way. The salmon lodges of Campbell River, including Ned Painter's and April Point, have attracted the wealthy and privileged from around the world since the early part of the century. Challenges like the Tyee Club rules are made to order for such extravagant excursions, implying as they do not only difficulty but ceremony.

To become a member of the Tyee Club even today, an angler must take a chinook salmon over thirty pounds — a tyee — while fishing from a rowed boat with line rated at a breaking strength of twenty pounds using only artificial lures. Period. You may have a guide who rows the boat, and you must land rather than release the fish to be awarded your membership button. Naturally, this is accompanied by festivities and formal good cheer reminiscent of earlier times. It is most definitely an activity apart from the feeding frenzy at Cape Mudge.

"Learning to row the pool," Alan Haig-Brown told me, "is one of those things where you learn seventy-five percent in the first hour, and then you take the rest of your life figuring out the other twenty-five percent. I got what amounted to the first lesson of a hundred, I suppose, and was never very good at it. I learned the lineups, like you steer toward the tree on the end of Orange Point, then line up with the second bump on that mountain, and then you follow that until you see the spruce tree, and so forth. All of that is why you need a guide, but the fish themselves are about as much of a thrill as anything you can do on the water.

"I suppose the appeal of the tyee is that it is massive, and also at the absolute peak of its development and ready to enter the river. They've put on their weight and beauty, and when you catch one you have come together with all of it," Alan said. Tyees over sixty pounds have been taken.

"What my father worried about, and what has happened is the pulp mill, and now they're building a marina right on the edge of the pool. The tyee aren't there as they used to be, though some are taken every year, but it's the overall change that bothers me. It would have bothered my father."

"A conservationist fights many battles," wrote Roderick Haig-Brown in *The Living Land*, "varying in scale all the way from the attempted protection of some individual species of wildlife to the supreme issue of proper use of soil, air, and water; and every fight is complicated, if not forced, by the false urgency and outdated sanctity of progress. . . . Timber, soil, fisheries, oil and minerals, even water power, become more, not less, valuable with delay."

The waterways and archipelagos from Puget Sound, along the inside and outside coasts of Vancouver Island, and into Southeast Alaska lend new meaning to the charm of messing around in boats. In many sections of this maritime overstatement, watercraft are essential to daily life, as on Bamfield Bay where the school boat, instead of the school bus, transports the precious cargo to and from daily lessons.

The San Juan Islands and their Canadian sisters, the Gulf Islands—a collection of low, forested hummocks on the boundary in Puget Sound—settled like hard-rock geologic tourists after the last ice age. The hardest thing about a sojourn through the San Juans, the saying goes, is leaving, though some sailors have found mornings like this one far too convincing to even try.

Part of the charm of the southern reaches of the Inland Passage, here in its ripple form, is delivered by the quaint but cosmopolitan harbor at Victoria, the capital of British Columbia. The city is a yachting center of the first order, a jumping-off place both for the leisurely meander up the coast as far as Southeast Alaska, or the hardcore North Pacific crossing to Hawaii. Annual race weeks in Victoria include the Swiftsure, which attracts international entrants.

The pleasures of Vancouver Island include its rugged west coast, the milder environs of the eastern shore, and access to mainland British Columbia, all of which can occupy a sailor, naturalist, or wanderer of any stripe. This meadow is near Kwai Lake on the Forbidden Plateau of British Columbia's magnificent Strathcona Provincial Park, which includes the island's highest peak, Mount Golden Hinde at 7,219 feet.

The now-dammed Campbell River and its tributaries emerge from the highest terrain on Vancouver Island near this meadow on the Forbidden Plateau. From the western watershed flow steeper rivers like the Gold, which heads in Nootka Sound, one of the dozens of spectacular intrusions into the wild ocean coast that provide dependable refuge for mariners and unsurpassed terrain for hikers.

The Pacific Rim National Park is a celebration of one of the world's most elegant mergers between land and sea. The park is a narrow preserve running along Vancouver Island's ocean coast from Port Renfrew on the south to Tofino on the north, a distance of almost a hundred miles. A well–tended trail offers the best kind of hospitality to hikers, and the coastline includes the safe harbors of Barkley Sound and Bamfield Bay.

The bald eagle makes its strongest showing in North America along the entire range of this book, from the Columbia River to Cook Inlet in Alaska. Though sightings are quite common, even absurd when the spawning salmon attract hundreds at a time to some rivers in the fall, an eagle always draws a glance and, frequently, a meditation.

The term *riparian* refers to habitat that occurs along the banks of streams and rivers, which are in no short supply on the Northwest Coast. Characterized by relative instability, most of the vegetation consists of fast-growing plants that can take advantage of short growing seasons and damp conditions. Among the first verdant arrivals after floods, glaciers, fire, or logging are the alders, seen here in winter light.

The remaining stands of old-growth fir are precious on Vancouver Island and throughout British Columbia, where virtually unrestrained logging practices have decimated the spruce, fir, and hemlock forests. These old fellows in Cathedral Grove, a special preserve, date back to the fifteenth or sixteenth century and participate in the carbon–water cycle essential to all life in ways we are just beginning to understand.

Vancouver Island's Kennedy River draining from the inland slopes of Mount Maitland to Barkley Sound is typical of the lake–river systems favored by salmon and other anadromous fish. The early aboriginal bands of seafarers used such waterways for access to food, timber for their canoes, and inland respite from the sometimes fierce coastal winters.

72

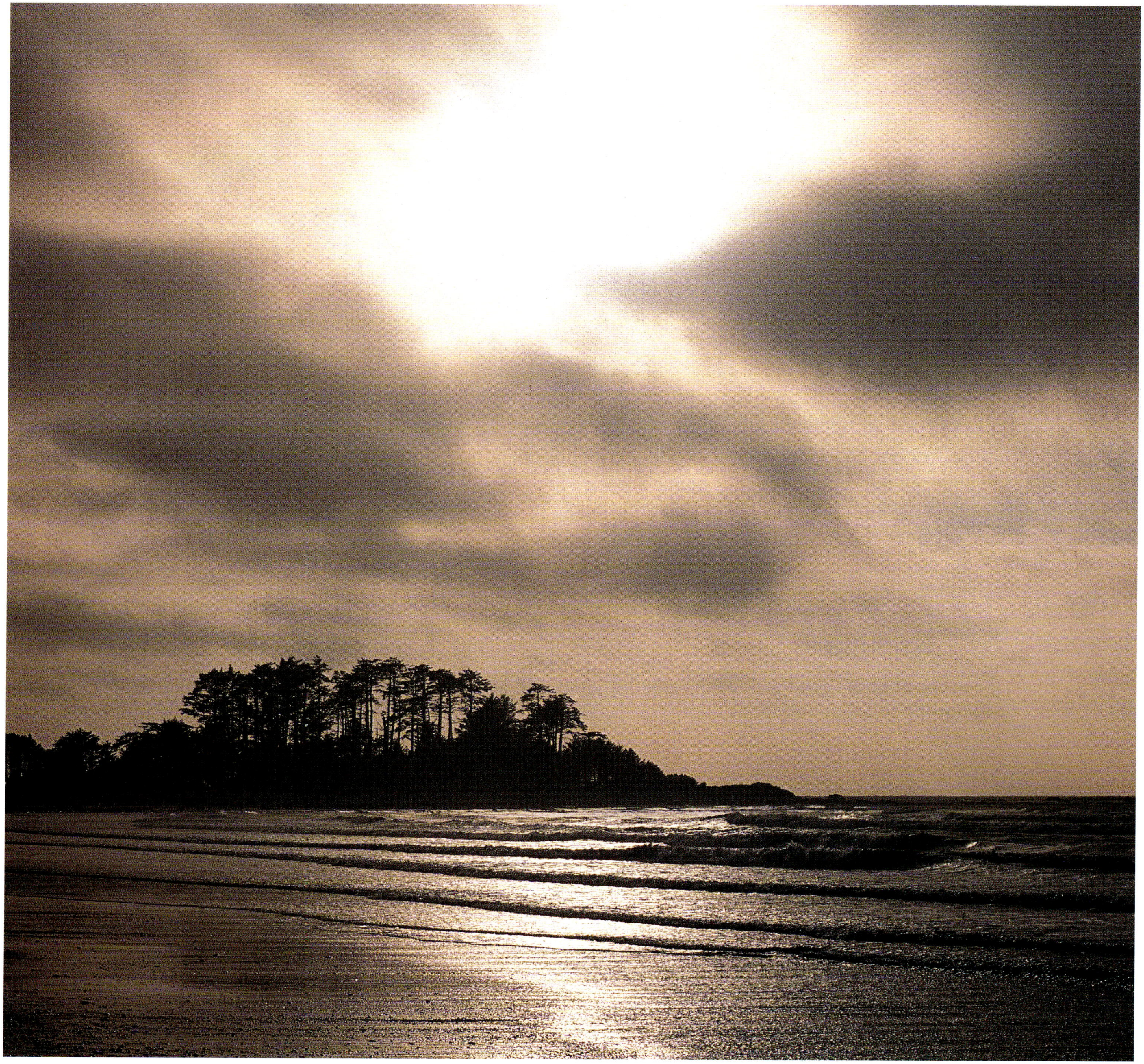

North Beach, on Graham Island in the Queen Charlotte Islands, British Columbia.

GOSPEL ISLAND

The log truck materialized in a thunderhead of its own dust like an avenger coming to claim us, rounding a curve with its headlights flashing, backlit by the afternoon sun to the west. The size of the thing alone was frightening, its radiator grill rising ten feet above the road, the roof of the dirty red-and-white cab with its protective cage adding another six feet. Towering into the blue sky above even the dust storm was the load of trees, sprouting branches that slapped the roadside brush in passing, giving the entire assemblage the appearance of an attacking tree-creature in one of those nursery books that had terrified me as a child. A forest log truck is twice the size of the biggest highway tractor trailer. This one, with its cargo of freshly cut, old-growth spruce and hemlock, weighed about 120 tons, almost a quarter of a million pounds, and it was rapidly devouring the gap to Ronnie Ross and me bouncing along in his Ford pickup.

"I hate trying to stay ahead of one of these guys until a pullout," Ronnie said, picking up his speed. "It's just a bad feeling to have it back there."

A pickup truck earned its living here in the Queen Charlotte Islands; this one was a light blue, four-wheel drive with both side mirrors broken and dents in the lower-body panels that were specific to travel over primitive roads. The truck was purely utilitarian, offering no amenities except a tape player, from whose speakers, laying loose on the dashboard, flowed the flute of Jethro Tull, Ronnie's favorite musician and fellow Scotchman.

The hard-worn Ford is Ronnie's primary tool. With it he hauls firewood to heat his house, garbage to the dump, his carpenter's kit to produce cash, and a boat for fishing. He migrated to the islands from his beginnings in Renfrew, an industrial scar near Glasgow, with a fifteen-year stop in Montreal to finish growing up, marry his grade-school sweetheart, Lorna, and realize that he needed a frontier to have a future. Two brothers preceded him to the Charlottes (as the islands are known to white locals), for the same reasons as everyone else who ends up sixty miles off the coast of British Columbia on the islands of the ancient Haida Indians, the westernmost outpost of the great Pacific rain forest. Now, at thirty-two as the last decade of the twentieth century begins, Ronnie is picking his way into midlife as a drywaller and nail-pounder, "looking for work and fishing," as he put it. He said he won't work for the logging company because selling his life by the hour doing something so dangerous and tedious seems a waste. He also said he hates what logging is doing to the forest.

We were towing a sixteen-foot skiff on a trailer, coming home to Queen Charlotte City from Rennell Sound on the west coast of this haunting, contradictory archipelago. It was a day trip with an hour's drive each way over the logging road through nonstop clearcut that looked more as if it were made by an asteroid strike than by ordinary mortals with chain saws. "This is really something," Ronnie said at one point when the vista was entirely hilltops of graying slash and crumbling slopes. "I can't believe it sometimes, and I don't understand why we don't seem to know some of the things about the forest that the Indians have known for a long time."

The Indians are the bands of Haida, seagoing migrants who sought the frontier of the misty island forests on the horizon. They came in canoes of exquisite and purposeful design, a fierce and spiritual streak of humanity given the privilege of ascendance during a time when natural resources were overly abundant. The Haida numbered perhaps thirty thousand at the time of the last ice age. They thrived on salmon, clams, mussels, and fruits of the forest such as berries and ferns and protected their sustenance in fierce warfare. To them the Charlottes are *Haida Gwaii*, "The Place of the People," and their myths tell of the first human being arising from a clamshell on the northern promontory now called Rose Spit.

The thriving bands built superb seagoing canoes from the giant trees and memorialized events, gods, and heroes in their architecture and totem poles. In art and subsistence, the Haida and the other aboriginal people of the Northwest Coast did not perceive nature as a collection of subordinate objects; for example, an ancient tree might physically exist apart from a human being, but no distinctions of essence gave either a superior status. Turning a tree into a canoe, therefore, was a spiritual matter, not one relegated to the pedestrian concerns of commerce. Even through the recent dissonance of abusive resource extraction, a vision of life in harmony is still possible in the Charlottes because of the raw power of so wild a place.

The Queen Charlottes are a cluster of 150 islands lying off the mainland between Vancouver and Prince of Wales Island. The geology of the Charlottes is an inspiring song of pressure and extrusion, suggesting a violent past, with the raw edges of tectonic collision producing granites and preglacial rocks uncommon elsewhere on the otherwise ice-formed coast. The last wave of ice ten thousand years ago nicked only the easternmost rim of the island group while covering and scouring the adjacent mainland. The natural history, flora, and fauna of the Charlottes are so different from those of the surrounding territory that they are often compared to the Galápagos group in the South Pacific.

The Haida and everyone else who followed them lived in the Charlottes in that state of superabundance until the mid-1940s, when the orgy of logging in British Columbia reached the islands. The clearcuts now are so vast they have been observed by astronauts in space, a sad fact made more so because the islands were virtually pristine until logging began. The forests of Haida Gwaii became another casualty of World War II when the timber companies arrived to take the Sitka spruce, which the airplane manufacturers used to construct the frames for their machines. Once that toehold was obtained, several companies, including Macmillan Bloedel on Graham Island and a company now called Fletcher Challenge on Moresby, secured licenses for cutting. British Columbia uses a fox-in-the-henhouse policy to harvest its forests, allowing the timber companies to fully manage the vast licensed acreage as well as cut the timber.

The ancient trees on the log truck bearing down on Ronnie Ross and me were probably bound for further reduction into disposable packaging like chewing gum wrappers, bags, and boxes, not lumber for houses, the Sunday edition of the *New York Times*, or some other trade-off we might be willing to make for the destruction of the Great Pacific Forest. And despite claims by the timber companies to the contrary, the new growth will not fulfill the water-bearing, carbon-replenishing, oxygen-generating functions of the old forest. We are just beginning to acknowledge that our understanding of intricate global ecodependency is shabby at best, and that we may already have robbed portions of the earth of their ability to sustain life as we know it.

In the early 1970s, the tragedy of the Queen Charlottes was revealed to the world when Haida bands on Moresby Island simply said, "No more." With encouragement and support from a new population of white people from Canada and the United States — the so-called hippies — the Haida blockaded logging roads on South Moresby, garnered international attention, and stopped the final stage in the destruction of the ancient forest. Now, under substantial Haida control, South Moresby is a national park with no logging or other commercial activity beyond carefully controlled expeditions. The other major park in the Charlottes is Naikoon, encompassing the shoreline and uplands surrounding Rose Spit.

Learning to live in the Charlottes, Ronnie and others told me, involves numbing yourself to the clearcuts and yielding to the astonishing beauty where you find it. Ronnie and I had spent the day on Rennell Sound, the largest inlet on the west coast, named after a geographer on the ship *Queen Charlotte*, which had given the archipelago its European name. Ronnie Ross, who had agreed to let me accompany him without hesitation after a chance meeting, wanted to try snorkeling for the abalone and large rock scallops he had heard were plentiful in Rennell Sound.

Ronnie is an instinctive naturalist with no formal training but a combination of curiosity and knowledge of the fauna and flora that makes him a good companion on such a trip. On a remote archipelago understanding nature means you can find food, and subsistence is a big part of life for everyone — white or Haida — in the Charlottes. On the way to Rennell Sound, Ronnie had told me that he is happy in the Charlottes, especially when he's gathering food, although he fears his wife's dissatisfaction with the isolation will eventually drive them back to a mainland city. "I'll tell you though, the first time I came over to Rennell Sound I went back and almost told Lorna, 'Lorna, we're staying, no matter what.'" Then he had told me of a special place called Gospel Island that had touched him particularly.

We had held back from heading straight for Gospel Island that morning because the best abalone gathering was reported to be nearer the head of the sound where we'd launched the boat, and in the unusual and hot sunny weather, there hadn't seemed to be any reason to hurry. The water was flat calm, a perfect mirror doubling the images of the surrounding, snow-streaked mountains that attain elevations of thirty-five hundred feet in that vicinity.

Ronnie struggled into a borrowed dry suit, mask, snorkel, and fins to explore the intertidal zone that lay exposed on an extremely low tide. He had never been in a dry suit or any other kind of diving gear before that day, but he gamely suffered the tight seals and suffocating heat before getting into the water, where the magnificence of what he saw turned his attention away from considerations of comfort. My chore was to follow him in the boat, paddling easily to stay a dozen feet behind, on guard but mesmerized by the splendor of this meeting between land and sea.

On the exposed granite shelf of the shore, eelgrass and kelp commanded a swimmer's first attentions. These fluid strands of oxygen-producing ocean plants offered not only sustenance of themselves and their colonies of minute animals, but protection to a universe of resident creatures. From the boat as I paddled, it seemed as if the sea were breathing as it soughed through the tidal circus, and like a big blue seal, Ronnie Ross kicked and paddled in the midst of it all. I could hear him exclaim through the snorkel when some great discovery found him, and the hollow grunt made me laugh each time. My eyes were often torn from the view below by other sounds and vistas that bespoke abundance and glorious variety. Birds filled their measures of the great song with their spectacle of voices and colors: an oyster catcher that looked as if someone had spent hours painting its thick, coral red beak to match its darting eyes; a large, brown sandpiper-like bird with a long, down-curving beak, obviously a tool evolved to extract food on a beach; and a massive bald eagle in a small pioneer spruce that bent slightly from the vertical under the bird's weight.

The snorkeling ended with a bucket of abalone, scallops, and sea cucumbers, all within the legal limits Ronnie had carefully memorized and to which he'd adhered. As he rested from his swim, chattering most of the time about the remarkable pantheon of life over the hum of the outboard, we made our way under bright skies toward the mouth of Rennell Sound and

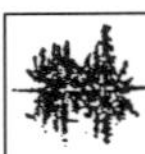

Gospel Island. A hospitable splash of white beach on the leeward side—the wind almost always blows from the west there—greeted our approach, and as soon as I set foot from the boat I sensed a power in the place.

Perhaps five acres in its entirety, Gospel Island is a miniature composite of the vital elements of the Queen Charlottes: old forest, rugged shoreline, and the press of the ocean's desire to overtake all of that. We dropped to the sand and basked, still slightly chilled from the wind of the boat ride and unable to proceed without luxuriating in the rare hot spring sun, and then we began to explore. Ronnie demonstrated again that he is a born naturalist and wilderness companion by allowing me to wander alone at first, giving a gift of solitude I had forgotten to request. I walked clockwise, intent on a circumambulation, and watched him disappear into the brush in the other direction.

My walk took me over granite ledges rolled like giant frozen pillows, covered in spots by lichen and amazing clusters of bright yellow flowers growing from the rock. The rock itself was scribed and fractured and in some places deeply cupped into pools in which sun-warmed water covered boulders that had apparently been deposited by the sea during storms. Ronnie joined me just as I came on a sculpted monolith facing the sea across a series of these pools, which looked as if they had provided baths for ancient travelers or had been the site of ceremonies inspired by exposure to the ocean's elemental forces. Both of us were struck by the sense that our wandering predecessors had also taken some message from Gospel Island that their spirits, rather than their rational minds, could grasp. We walked on, but spoke in muted tones as though participants in some solemn rite.

I raised my camera to the scene over the chain of pools, focused to include the blue-black ocean and the variations of color on the hills surrounding Rennell Sound, the deep, almost purple tones of the trees on the steepest parts of the hills, the lighter greens of emerging trees where logging had taken the old growth, and the brownish scars on the hills where the trees had been skidded down to the beach—and realized I was out of film and had brought none with me from our beach camp. I muttered that this was a piece of bad luck, and Ronnie, standing within earshot, replied, "I guess the rest is just for the mind, then."

We circled Gospel Island, were harassed by a pair of oyster catchers decoying us from their nest, crept over a ledge to peer into a pocket cove in which a colony of seals had encamped for the breeding season, watched guillemots in their mating overtures, and then we left. Two hours later, after loading the boat back on the trailer and reentering the clearcut zone between Rennell Sound and Queen Charlotte City, we raced the log truck to a pullout. For several minutes, it was barely a car-length behind us, close enough so I could read its manufacturer's name—Pacific—through the back window of the pickup. Finally, we scrambled off the right-of-way, and the monster rumbled by, enveloping us in its sheath of dust and gravel. Ronnie Ross turned up the volume on Jethro Tull, looked over at me, and grinned. "I still like it here," he yelled.

The earliest known settlers in the Queen Charlotte Islands were the Haida, who called the archipelago *Haida Gwaii*, or "The Place of the People." In the Haida creation myth, human beings rose from a clamshell on Rose Spit, at the northern big island. Likely, though, the Haida arrived sometime soon after the retreat of the ice, about ten thousand years ago. Europeans, like Sister Mary whose grave is on South Moresby Island, didn't settle until just a hundred years ago.

Mariners on the Northwest Coast have been running past the Queen Charlottes at least since Juan Perez recorded landfall there in the mid-eighteenth century. Soon after, Captain George Dixon, an Englishman, approached the islands from the west, naming them after his ship, the *Queen Charlotte*. Both Perez and Dixon, of course, were preceded by the great Haida seafarers, who carved their craft from whole giant spruce logs.

If you are one of the six thousand permanent residents of the Queen Charlottes now, a weekend on the town means Prince Rupert, a ferry ride of several hours across Hecate Strait. Prince Rupert is the hub of the northern British Columbia coast and, like its outlying partners to the south and across in the Charlottes, depends for its existence on resource extraction—logging, mining, and fishing.

The high mountain ranges of British Columbia extract tremendous amounts of moisture from the storms that come in from the Pacific Ocean. Roughly one-fifth of the rainfall over the land drains back to the ocean through the creeks and rivers. The balance is stored in snow and ice of glaciers and snowfields. This lush scene is Englishman River Falls.

Coastal British Columbia is made up of steep mountains. It is part of the Cordilleran System, the great chain of mountains that run through the Americas from Point Barrow to Cape Horn. The coast range receives rainfall ranging up to 150 inches at the higher elevations. These conditions create waterfalls of endless variety and beauty like Shannon Falls shown here cascading over the rocky cliffs.

All forms of life on earth depend for their existence on their water. In a complex but predictable pattern of climate owing to temperature, gravity, and the velocity of our rotating planet, water rises as a gas from the sea and land, condenses into liquid in the atmosphere, falls as rain, fog, hail, and snow, and returns to the seas. This is called the hydrologic cycle. The amount of water participating in the earth's hydrologic cycle has remained relatively constant since the last retreat of the continental glaciers.

About 99 percent of the water on earth is in the oceans and polar caps, with the remainder in rivers, glaciers, and groundwater, and at any given moment a small fraction is distributed through the atmosphere in the form of vapor. Garibaldi Lake, in Garibaldi Provincial Park on the British Columbia mainland, is shown here in morning light doing its part.

The fairly typical flora and fauna throughout the territory we call the Northwest Coast owes its consistency to the revival of life since the last glacial retreat, about ten to fourteen thousand years ago. The glacial scouring apparently was a kind of pruning, and when the ice was gone, the hardy forests transformed the earth here into a paradise in harmony with water.

Reid Inlet, Glacier Bay, Alaska.

DIXON ENTRANCE TO CAPE SPENCER

Chief Kadashan Totem, Wrangell, Alaska.

THE SWIMMERS

Approached at water level, the thing looks like a raft of drifting logs and debris on which some misguided adventurer has built a shack. A lattice of boards, ropes, and cables emerges as you draw near, but still it begs to be described as a contraption, or some seagoing mistake. From the air, though, if you happen to gain such a view, a pattern emerges that reveals the purpose of this floating puzzle; once you realize what it is, an ingenuity that draws on centuries of survival is obvious.

Against the jade green water, the wooden structure covers a rectangle of about 200 by 150 feet with mesh descending beneath the surface — clearly a submerged cage of some kind. Catwalks of light-colored boards are fixed over logs; anchor lines under strain against a running tide extend from several points along the perimeter. One end of the rectangle is closed off like a box; the other end is open, with the logs and boards forming a series of V-shapes that lead into the closed end. If you zoom in for a medium close-up from the overhead view, the riddle is solved conclusively by the flashing silvery shapes in what is obviously a fish trap.

Before Alaska became a state in 1959, fish traps were the mainstay of a booming salmon-packing industry, and just about as common as cruise ships are now on the waters of the Inside Passage. Fired up by anticolonial outrage and pressure from independent fishermen, though, the very first Alaska legislature banned all fish traps, most of which belonged to packing companies from Washington, Oregon, and California. Now, only four traps exist, owned by a cooperative of Tsimshian Indians from the village of Metlakatla on the west side of Annette Island, the only Indian reservation in Alaska. Reservations are essentially federal enclaves within state borders, and many concessions have been made to their residents to allow self-sufficiency. Bingo, fireworks, and tax-free cigarette sales are familiar examples; here on Annette Island, it was fish traps.

The Tsimshian came to Southeast Alaska from the Skeena River region of mainland British Columbia, led by an unordained Anglican missionary named William Duncan. En route, during what proved to be a migration that spanned fifteen years, Duncan and his flock established their first utopian Christian community, called Old Metlakatla, near Prince Rupert, in 1862. Their pri-

mary industry was catching and packing salmon both for sale and subsistence.

Duncan, a bit of a renegade free-thinker, was in constant conflict with his superiors in England. In 1887, having gained the ear of American sympathizers in the Alaska Territory just over the border from Prince Rupert, Duncan and about two hundred of his people moved to Annette Island and founded another village they again named Metlakatla, also based on salmon packing and utopian ideals. After Duncan's prolonged and eventually successful lobbying job on the U.S. Congress, Annette Island was designated a reservation, really a sanctuary of sorts, under the oversight of the Bureau of Indian Affairs.

Most Metlakatlans are still Tsimshian, one of the three aboriginal strains of people who settled the islands and coastline of the Pacific Northwest. Southeast Alaska, sometimes called the Panhandle, is an archipelago of more than a thousand islands occupying a strip eight hundred miles long by about one hundred miles wide between Dixon Entrance and Cape Spencer. Southern Southeast Alaska, where Metlakatla is located, is quite different from the northern zone, with generally lower mountains, more timber, and more rain. Hence, more salmon. The Tlingit (pronounced Klink-it) and Haida bands probably came into the archipelago from the south and east, moving through the coastal mountains along great rivers: the Skeena, the Stikine, and the Taku.

The Tlingit, Haida, and Tsimshian people shared similar social traditions, and were organized in clans allied with specific animal totems, such as Raven, Killer Whale, Wolf, Eagle, and Bear. Family and clan rights passed from the mother's side, and clans claimed and defended territory and, most significantly, fishing grounds. The great works of monumental art we call totem poles marked territory. Many remain in villages in Southeast Alaska, along with the historical perspective of the aboriginal culture that placed humans in, rather than apart from, the rest of the animal world. The ceremonies of eating, hunting, and gathering preoccupied the ancient clans of the Northwest Coast and accounted for the major spiritual threads woven into their tapestry of life.

White European settlers began their own migrations to Southeast Alaska in the middle of the eighteenth century when Vitus Bering opened up what would be called Russian America. This was an impulse of the fur trade, which eventually centered in central and western Alaska and away from the capital of Russian America at Sitka in the outermost island group of the archipelago. French fur traders from the mainland of British Columbia, agents primarily of the Hudson's Bay Company, and other voyagers of that early era of white settlement encountered Southeast Alaska while searching for routes to and from other places. The miners who came for the Klondike gold also used the Inside Passage for safe transit, but few of them stayed. Not until the late nineteenth century when the white people came for the salmon did their settlement of Southeast Alaska begin in earnest.

Modern Metlakatla is Alaska's southernmost city, a going concern of about fifteen hundred souls, set in a verdant paradise of spruce, hemlock, harbors, fertile coastline, lakes, and streams that seasonally host salmon. Until Ketchikan, on a bigger island just to the north, built its own runway in the early 1970s, all air traffic into southern Southeast Alaska came through a Coast Guard airfield near Metlakatla. From there, passengers continued to Ketchikan, the largest city in the region, by float plane.

Nothing influenced the heritage of both the aboriginal and white settlers of the American North Pacific more powerfully than salmon, probably known to the early inhabitants as simply "swimmers." As a symbol, the Pacific salmon appeals to our most obvious and profound capacities for wonder: With a certainty both elegant and dreadful, the salmon hatches in freshwater, migrates to the sea, endures threat and hardship, feeds until sexual maturity, miraculously returns to the stream of its birth, then spawns and dies. From May to October, the watercourses of the Pacific Northwest, ranging from the Yukon, Fraser, and Columbia rivers to the most meager seasonal capillaries, receive the salmon.

Although everywhere in the drama of existence birth implies death, its occurrence in the visible spawning cycle of salmon seems to reveal such truth more passionately than in other settings. What's more, nature's most elemental paradox — that one organism must die for another to live—is also fully portrayed in the salmon's turn on earth. The abundance of the North Pacific Coast depends almost entirely on the annual rhythms of the salmon, with seasonal feasts for birds, mammals, and sea life fixed on the salmon's emergence and return.

Everything is food. In a wonderful twist, salmon feed on other creatures far from where they become food themselves. They bear a huge share of the burden of nourishing the Pacific Northwest because they are able to leave it, feed in the deep and fertile gyre of the central ocean hundreds of miles away, and bring food back to the land in the form of their own bodies. It would be hard to imagine a more perfect animal to perform such a valuable function, that is, moving nutrients around in a complex system. The salmon are nitrogen-bearers, just up the line from the pioneer plants like fireweed and lupin that scratch out that life-giving element from the broken rock left in the wake of glacial retreat. Their roots break up and nourish soil, streambeds, and the necessary gravel redds for spawning salmon, which in turn nourish the earth with their bodies. The successions of organisms and life cycles seem to fit together so perfectly it is difficult to keep in mind the ultimate randomness of it all.

Salmon are rich in the subtleties of evolution, and the empirical compulsions of the European settlers of the North Pacific were not served by the classification of all salmon as simply swimmers. The physical differences in the five members of the genus *Oncorhynchus* probably derive from ancestral variations in their time of return, available nourishment, length of the rivers of origin, and seasonal anomalies during their genetic progress. Salmon adapt so quickly, in just a few generations, they have become the perfect vertebrate for genetic research.

The Greek root of the word *Oncorhynchus* means "hook nose," hence the genus under which Pacific salmon have been classified since the early Russian excursions into the territory. In descending order by size range and relative modern market value, the five members of the genus, along with their regional names, are *Oncorhynchus tshawytscha* (king, chinook, blackmouth, tyee, spring), *O. kisutch* (coho, silver), *O. nerka* (sockeye, red), *O. keta* (chum, dog, calico), and *O. gorbuscha* (pink, humpbacked, humpie). A sixth member of the genus, *O. masou* (cherry salmon), spawns only in Asian Pacific waters.

Atlantic salmon are classified with seagoing trout, genus *Salmo*, although they are probably related to the Pacific salmon in the not-too-distant geologic past. Their cousins are the steelhead trout. In 1988, fishermen on the Pacific began occasionally catching Atlantic salmon that had escaped from marine salmon farms, net-pen complexes in which the fish are fed and raised in captivity from fry to harvest. The exact effect of this on the North Pacific food web and salmon gene pool has yet to be determined, although the farms of British Columbia and Puget Sound are clearly proliferating.

As recently as the last century, Atlantic salmon ascended the great rivers of North America and Europe in such numbers that farmers reportedly pitchforked them from the banks for fertilizer. The salmon spawned in tributaries of the Rhine, for instance, as far inland as Switzerland. Now, pollution and the other ill effects of human settlement have so decimated wild stocks of Atlantic salmon that they are no longer a food resource.

The ancient settlers thrived in Southeast Alaska and the rest of the Pacific Northwest because, without knowing the details, they were served by the salmon's ability to bring food to them. The Indians figured out how to build weirs or traps of rocks, wood, and sinew across streams to harvest, and later smoke, a winter's supply of salmon. The other animals that came to share in the feast became food as well. The Indians also learned to fish with hooks and nets for halibut, rockfish, and herring, all of which were extremely abundant until just a hundred years ago. The earliest white fishermen packed their salmon out, salted in barrels, in modest quantities, or took the fish seasonally to sustain themselves just as the locals did.

Traps are an obvious way to catch salmon since they return to spawn predictably in their home streams. Chasing them around with boats can be considered an expensive romantic activity. Boats have been used since the turn of the century but accounted for less than half the annual pack until the traps were banned. A salmon trap is nothing more than a box of netting set in the path of migrating salmon; strategically placed barriers, called leads,

direct the fish into the holding pen, or heart of the trap.

Traps work especially well for salmon because they are reluctant to reverse course to swim with the current. This is a primal instinct since they must swim upstream to spawn. No specific design exists for a salmon trap, although leads and pens are generally present, and traps may either be floating or fixed. The highly efficient traps introduced by the white men are not aboriginal in design, but they do harken to the ancient methods of simply barring passage to gather the fish for food.

The traps at Metlakatla are typical of the floating type, which were most common in the deep water of Southeast Alaska where traps suspended from pilings driven into the bottom of the rivers or streams were impractical. Whether fixed or floating, traps were expensive and, during the season, required constant repair, usually performed by crews from the cannery. A watchman lived aboard — hence, the shack — to drive off trap pirates, sharks, and marine mammals that threatened the catch.

Had the traps been more carefully regulated and avoided conflict with fishermen, they might have remained the mainstay of the salmon industry. However, their efficiency, as much as anything else, became a major conservation problem. When the process of canning was perfected in the nineteenth century, the salmon returning to the Pacific Northwest were feeding the people of an entire world, not just the local population. The pressure of competition on such a scale encouraged abuse.

Salmon were perfect for canning, if the process was performed correctly. Until they expire at the ends of their spawning runs upriver, salmon are firm-fleshed, oily fish whose bones literally dissolve under heat and pressure. With the addition of a touch of salt, careful sealing and retorting, nothing else needs to be done to preserve a good-tasting piece of protein. The technology alone enabled a few packing companies to establish hegemony over the salmon of Alaska, and eventually the people of the territory threw off the yoke and banned the salmon trap.

The first salmon to end the animated part of its life in a can did so in 1864 on California's Sacramento River on a barge owned by John, Joseph, and R. D. Hume and their partner, Percy Woodson. The first canneries in Alaska were built in 1878 in the villages of Klawock and Sitka, not too far from Metlakatla. By the turn of the century, hundreds of canneries were puffing away in the Pacific Northwest from the Columbia River to the Bering Sea. They numbered in the thousands a decade later, and then, not surprisingly, the salmon began to disappear. Unrestrained, the packing companies were grave destroyers driven by a hungry world. Markets sucked up as much salmon as the companies could produce.

Nature's truth holds, though: We have to kill to live. And the more recent devastation visited on the salmon by dams, pollution, and habitat destruction far outstrips that inflicted by the canners' capitalism. The population of the world will double to over 11 billion by the year 2030. Some futurists say traps will again bloom at the mouths of streams when fossil fuel depletion signals the end to fishing fleets, and fish farms will provide the world with an abundance of protein to match that of the ancient settlers of the Pacific Northwest. The Tsimshian of Annette Island, then, would have an edge.

Dixon Entrance to Cape Spencer

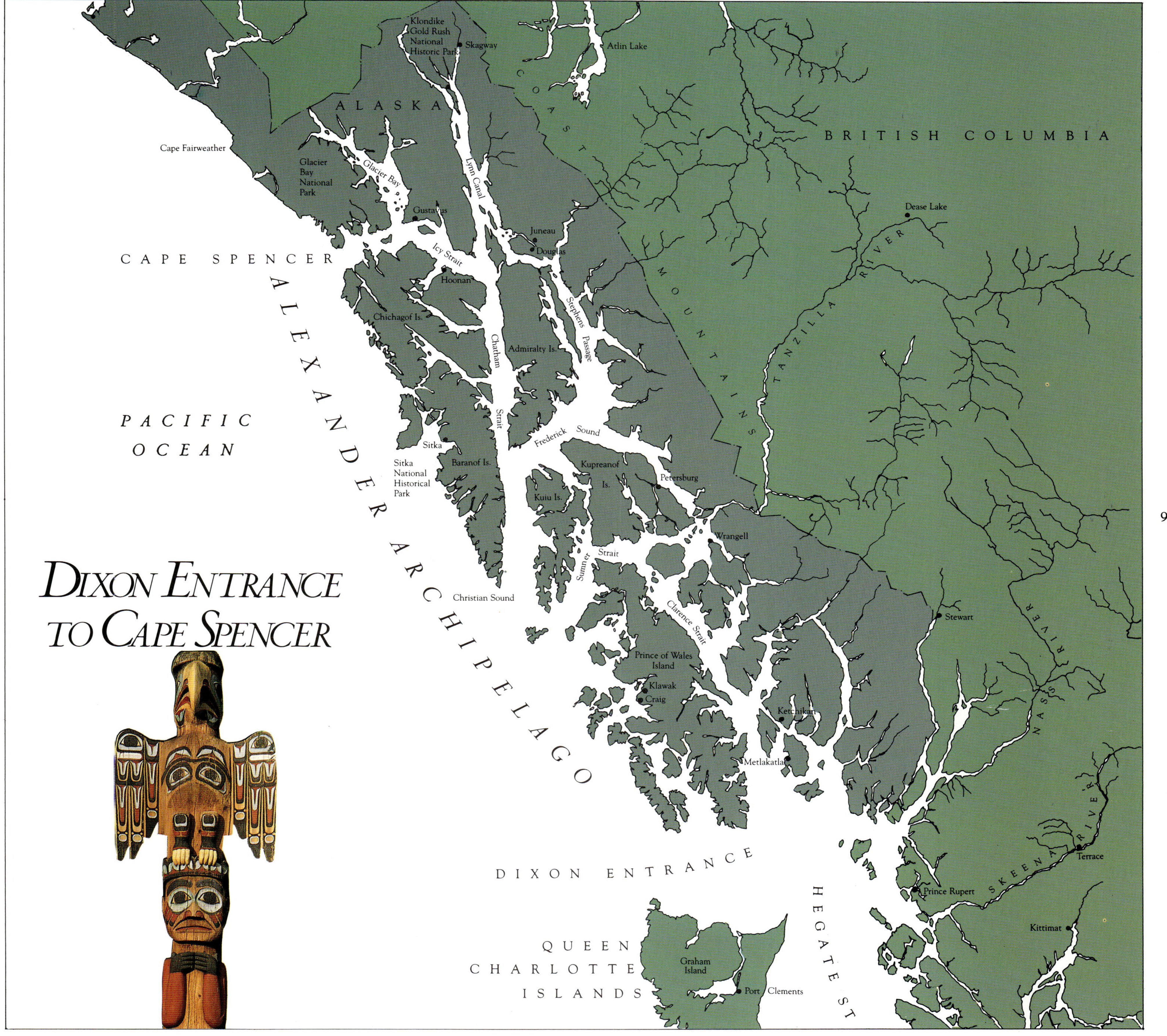

93

Landfall in Alaska for most travelers from the south has been Ketchikan, now a town, once a village, and for thousands of years before that a place the Tlingit seafarers called "the salmon creek that flows through town." The blend of European and Tlingit cultures has produced anomalies like the Chief Kyan totem juxtaposed against the Victorian architecture. Legend has it that anyone touching this totem—not an unpopular gesture among the usually adventurous new arrivals in Ketchikan—will have money within twenty-four hours. Oddly, the Tlingit had no use for money at all, conducting their commercial affairs with gifts instead.

It would seem that everything in Alaska is connected to the water. From Wrangell, you can voyage to the thousands of islands of the Alexander Archipelago. This is arguably the most exotic and hospitable cruising ground for the wandering mariner. Lush bellies of water like Frederick Sound and Icy, Sumner, and Chatham straits are laced together by navigable but astounding narrows and passages, all in the embrace of the mountainous rain forest broken in places by estuaries fairly vibrating with life, river outfalls, and tidewater glaciers.

In the summer, the good harbors at Ketchikan, Wrangell, Petersburg, Juneau, Haines, and Sitka bring ferries and cruise ships like enormous migrating animals. Villages like Craig, Angoon, Hoonah, Pelican, and Elfin Cover conceal themselves a bit more than the mainline towns on the Inside Passage and in short order can bring new meaning to the notion of human settlement in the wilderness.

The Stikine River flows from the heights of the Canadian Coast Range, skirts the southern edge of the LeConte Icefield, with peaks including Devil's Thumb, Agassiz, and Kates Needle, and settles into the waters of the archipelago, dispensing its nourishment and debris in a great delta in the vicinity of the town of Petersburg. More than a few adventurers have journeyed to the Stikine–LeConte Wilderness Area for a vacation kayak expedition and simply decided to stay in Southeast Alaska.

The mainland icefields like those in the Stikine–LeConte Wilderness Area are the last remnants of the ice age. The glaciers that mark the edges and lower elevations of the fields are studies in pressure in which annual snowpack and warming conditions govern advance or retreat. The patterns of individual glaciers tend to remain constant over hundreds of years—some advancing, some retreating—but overall the volume of ice is still in decline.

Crab pots (like these stored near Petersburg), nets, and tubs of hooks and line are everywhere in Southeast Alaska, testifying not only to the commerce of the fishermen but to the abundance of the sea nearby. The list of plentiful species on which both local and seasonal fishermen feast is astonishing: five types of salmon, halibut, black cod, rockfish, herring, Pacific cod, lingcod, Dungeness crab, king crab, sea cucumber, sea urchin, scallops, shrimp, and abalone.

During the time of superabundance, until about 150 years ago, the aboriginal settlers including the Tlingit, Haida, and Tsimshian wove themselves into the food web with an intricate blend of efficiency and spirituality. Modern European fishermen gained their toehold in the salmon fisheries of Southeast Alaska in the late 1900s, when they came for brief seasons, salted their catch, and left. The tin can changed all that, and by the turn of the century, industrial food gathering had forever altered the place.

People talk about Sitka as being "outside," meaning out on the exposed Pacific coast of Chichagof Island on the western edge of the archipelago. The dominant fishermen are the trollers, who fish on the open ocean for the world's highest-quality salmon with hook and line from small boats distinguished by their long outrigger poles from which the lines are trailed.

From 1804 until the United States purchased Alaska in 1867, Sitka was the Russian capital in the New World. Though the town now celebrates that time with folk dancers, and festivals, the era of the Russian fur trade, in truth, was an orgy of commerce marked by cruelty to the aboriginal people and little concern for the animals or habitat beyond profit. The fur traders, or *promyshlenki*, made their own rules with civilization and the tsar so far away. This is a view of the preserved St. Michael's Russian Orthodox Cathedral.

The Resource Institute's schooner *Crusader*, in Windfall Bay on Admiralty Island, Alaska.

A BEAR STORY

In the mid-1970s, the citizens of Juneau commissioned a sculpture for the plaza of the new state courthouse, an institutional gray tower set in the center of town. The building was the latest in an invasion of large public edifices financed with petrodollars from the pipeline, and mixed emotions were the drink of the era. Everyone had a job, but big government had arrived like a rich relative with very bad taste. A piece of fine art, the townspeople decided, might improve this slab-sided new courthouse. So a committee chose the artist, a well-intentioned fellow from "out-of-state," which was like a racial slur at the time. The sculpture was executed and installed on the courthouse plaza.

It was an abstract piece called *Nimbus*, a twenty-foot-high spiral of welded steel painted a shade of aquamarine that nobody had ever seen before. At the unveiling, the most polite response was dead silence; many people made rude comments out loud about the artist's ancestry. The outrage continued for two years in the Letters-to-the-Editor section of the local newspaper, which is the living room of any small town. The few pro-*Nimbus* letters were thoughtful discourses on freedom of expression, not vindications of the sculpture at hand, which was unanimously considered a monstrosity. Finally, *Nimbus* was removed and carted off to who-knows-where, and flag poles were installed on the spot where the sculpture had stood like a mass hallucination. (I've always felt sorry for the artist who must have worked under the burden of some horrible misunderstanding about his clients.)

Ten years passed before anyone had the gumption to suggest putting another sculpture on the courthouse plaza, but everyone loves this one: a bronze, life-size brown bear, reclining as though after a good meal on a warm day. People decorate the bear with ribbons on holidays, stand near it and trace the contours of its metallic fur, and caress its massive head as they walk by. Created by Skip Wallen, a local artist with impeccable credentials as a wilderness observer, it is the most perfect rendition of a bear anywhere. The bear is well-suited to the reflections appropriate to a courthouse plaza, and it rings with the spirits of the ancient people of Southeast Alaska who knew the power of a god when they saw it.

During the years between *Nimbus* and Skip Wallen's bear, a similar though far more consequential drama of local outrage was played in Juneau and on neighboring Admiralty Island. This drama also included bears, and turned again on the tradition of populist uprisings that embroiders so strong a sense of community into the fabric of Alaskan life. Many, however, regard the outcome as among the most profound gestures we have ever made to acknowledge our dependence on the environment for something other than cash: In 1980, the primal forest of Admiralty was snatched from destruction and the entire island designated a national wilderness area, protected forever from commercial exploitation.

Admiralty is enormous, just over a million acres, with stands of climax forest never intruded on by significant human settlement. The second largest island in the Alexander Archipelago, Admiralty is ninety-six miles long and thirty miles wide at its widest, with about seven hundred miles of coastline tracing many bays, coves, and inlets. Chatham Strait, Stephens Passage, and Frederick Sound embrace Admiralty on the west, east, and south, and Admiralty's northern promontory, Point Retreat, gives way to Lynn Canal.

The island's geography is further defined by many large lakes and streams and a central mountain range rising to peaks of almost five thousand feet. The climate is distinctly soggy, with over a hundred inches of rain each year to nourish an ecosystem that is more water than not. The coastal bays and estuaries support grasses and floating populations of kelp, along with the pioneer efforts of lupin, skunk cabbage, and the other fast-growing plants that depend on quick response to short summers.

Inland foliage over much of Admiralty consists of pristine stands of spruce and hemlock, uncut forests nourished in stages by their decaying predecessors. These monarchs are attended in their mute journey through the ages by seasonal understory of alder, grasses, small flowering plants, salmonberries, devil's club in astonishing numbers with their prickly stalks and broad, sun-hungry leaves, and sphagnum moss in some places reaching depths of more than fifteen feet. Resting by a trail on such a fragrant moss cushion is a never-to-be-forgotten joy.

The tree line on Admiralty is at about two thousand feet, above which an alpine belt of mountain hemlock, lodgepole pine, heath, sedges, crowberry, and blueberries gives way to broad meadows of scrub vegetation. The vistas are comprehensible only in superlatives (and long-winded description), served as they are by the infinite greens, yellows, and browns of the forests and meadows, and the often-visible tidewater that reaches from dense blue to the chrome spectacle of the reflected subarctic sun.

Admiralty and the rest of the archipelago assumed their present form during the most recent glaciation of the Pleistocene epoch when a six-thousand-foot-thick ice sheet covered the entire region. A warming cycle began about ten thousand years ago, not only baring scoured earth and rock, but allowing the landmass to rise as the weight of the ice was removed. For awhile, the remaining ice shield retained a substantial portion of the earth's moisture, leaving the oceans at their low ice-age levels. So, there was a time when Southeast Alaska was not inundated with water, when land bridges allowed animal migration to what became islands. And during this time, from the remnants of their species scattered in a survival retreat to the ice-free parts of the continent, the great bears came to Admiralty Island and its neighbors Baranof, Chichagof, and Kruzof.

Ursus arctos, also known as the brown, grizzly, or Kodiak bear, is the principal inhabitant of Admiralty Island, a dominant mammal celebrated worldwide in myths and saloon yarns. Like humans, bears are omnivorous, thriving in a habitat rich in salmon, berries, roots, and other mammals from the Rocky Mountains to the Aleutian Chain. From one thousand to fourteen hundred brown bears live on Admiralty — the largest concentration on earth.

The rulers of Admiralty, characterized by dark coats and sizes in the middle range of their species, are known as the Shiras variety of grizzly or brown bear. They are immune to competition from animals other than their own kind. Black bears, common almost everywhere else in the hemisphere, do not live on Admiralty — nor do moose, mountain goats, foxes, wolves, porcupines, squirrels, or rabbits. Black-tailed deer, which probably arrived by swimming rather than by land bridge, thrive on the island. The only apex predator to share the island with the great

bears (and a few humans) are several thousand bald eagles in another of nature's greatest concentrations.

Bears are canny hunters, but their dietary preference runs to 90 percent vegetarian. They roam and forage in predictable patterns; some bear trails may be hundreds of years old, and the animals themselves live for thirty years or longer. Breeding takes place in midsummer, and after a 225-day gestation period, cubs are usually born in pairs, weighing about a pound each. Hibernation, the bears' most familiar biological trait, is a metabolic hiatus that begins in October or November when they seek the safety of dens to sleep through the winter.

About sixty thousand members of *Ursus arctos* remain in North America, two-thirds in Alaska and most of the rest in the Canadian north, with a remnant population of about a thousand scattered around the U.S. Rockies and a few reported in the North Cascades. Once there were hundreds of thousands; their demise is attributable to habitat destruction by human incursion and to the mutual fear between bears and humans in which the humans have dominated. Bears are dangerous animals; they do attack humans to defend their territory or young. And although everybody in the northern forests has a bear attack story to tell, humans usually win. Humans also hunt bears for sport, an activity unique to our species; few of us eat bear meat to subsist.

The ancient human settlers of the forest paradise of Admiralty were the Tlingit, seafaring people who came into the archipelago from the mainland soon after the ice withdrew. Living in patterns similar to those of the bears, the Tlingit's seasonal and permanent villages occupied food-gathering sites that were linked to the annual returns of the salmon and to autumn harvests of berries and other vegetation. The salmon also brought seals, sea lions, and other mammals, which the Tlingit dominated for the most part as well. The great bears, though, were another matter entirely; they were willing to defend their claims against all comers including human beings. Generally, the bears were ceded the interior of the island and the humans took the coast, though harmony did not always maintain — especially at salmon time. Local legends recount frightening encounters with the giants of Admiralty's interior.

The remaining Tlingit of Admiralty live in the village of Angoon, an enclave of about four hundred on a spit at the entrance to Mitchell Bay, the main interruption in the island's west coast. They regard the bears as supernatural beings and call the island *Hutsnuwu*, or "Fortress of the Bears." Since the end of the last century, white settlers, including prospectors, fish packers, and whalers, have trickled to the island. For fifty years, dozens of fish traps have pocked the coastline feeding canneries in all the bays and coves, but no permanent white towns took root on Admiralty.

A few homesteaders settled on the island, most notably a lone traveler named Allen Hasselborg, who lived as a hermit at Mole Harbor for thirty-five years. Hasselborg guided hunters and naturalists until his death in 1956 and probably knew more about bears than anyone in the world. Another homesteader was Stan Price, who lived on the northern end of the island at Pack Creek in pleasant coexistence with the bears until he died in Juneau at the end of 1989. Unlike the hermit Hasselborg, Stan was a generous man with a flair for easy hospitality. He created and oversaw the premier bear observation and research site, which is open to the public by arrangement.

Other than the villagers of Angoon and a few denizens like Allen Hasselborg and Stan Price, though, Admiralty belonged to the bears who were unchallenged in any significant way until 1968. Then came the largest timber sale in history from the federally owned Tongass National Forest, of which Admiralty is a part. As one of the buyers, Champion International got set to build roads and clearcut about 20 percent of the island, which would forever alter Admiralty's entire ecosystem. Until that point, less than a thousand acres of the island had been logged, making its forests among the few uncut wildernesses in the Northern Hemisphere.

The threat to Admiralty posed by the Champion sale galvanized an unlikely coalition of supporters, including Stan Price, hunting guide Karl Lane, Juneau lawyer-turned-filmmaker Joel Bennett, the full national force of the Sierra Club, and hundreds of local citizens. Some say Admiralty became a *cause célèbre* because it is just twenty air miles from the capital of Alaska.

Many people had trekked and hunted there, claiming shares of its majesty and thus some responsibility for it as well. Some say, though, that the bear spirit just wouldn't tolerate the transformation of the island into a scarred parody of itself.

Because of the island's unique pristine state, the Sierra Club sued to stop the Champion sale as soon as it was announced. A court issued a temporary injunction, but only an act of Congress could permanently vacate the sale and set Admiralty aside in perpetuity. At the time, Joel Bennett was working for the state but spending a lot of his spare time on the island taking pictures and developing a deep friendship with Stan Price. "The dominant spirit of the island is the brown bear," Bennett recalls of his fascination with Admiralty. "It is obviously their place, and it occurred to me that we needed to get the inestimable treasures that exist there on film. I was taking mostly stills at the time but decided to learn how to work in motion pictures because I could reach a wider audience."

He went to work on a film, *Fortress of the Bears*, financing it himself until the Sierra Club saw some of the footage and realized its power. Eventually, Joel Bennett's movie aired on national television and was shown to Congress, opening many minds to the value of pristine wilderness without a dollar sign on it. Bennett has been a filmmaker ever since, working mostly for a British nature series, but that first effort was one of the two most consequential turns in the battle to save Admiralty.

The other was Karl Lane's public opposition to the sale. Lane had known Allen Hasselborg, taken over his homestead at Mole Harbor, and tended the knowledge of the great bears during his own lifetime. (Lane died in a snorkeling accident in Hawaii in 1986.) "The fat hit the fire when Karl found out that Champion was going to build a low-grade pulp mill and ship the stuff to Japan," remembers his wife, Pat. "It was tough to talk him into siding with the Sierra Club because he didn't always agree with them, but he just stood to lose too much and knew everybody else would, too."

Karl Lane brought further revelation of the need to set Admiralty aside. His perspective evoked empathy for the bears, a surprising turn because usually stories about the great animals depend on fear and self-defense, on seeing brown bears as monsters, as an enemy to be feared rather than protected. But here was a hunter fighting to save the animals he hunted. People realized something special must be up with this woodsman and the bears. Karl Lane probably brought about the deaths of seventy-five or a hundred bears on Admiralty during his lifetime, but he saved many times that number by helping preserve the island wilderness. (Skip Wallen, who created the bear on the courthouse plaza, often traveled on Admiralty with Karl Lane.)

"It was all or nothing, because logging had really gotten out of control in Southeast Alaska," Joel Bennett says. "People finally stood up and said, 'We must have places that are not significantly altered by commerce.' The people said no. Enough is enough."

The settlement was a complicated piece of business that dragged on for years, involving the courts, Congress, and even the president. Finally in 1980, Admiralty was declared a national wilderness area, our highest protective designation. Champion was compensated but had to go elsewhere for its pulp timber; the Tlingit of Angoon received title to land around their village. And anyone can still go to the Fortress of the Bears and know such a place.

Admiralty was known to the ancient people as *Hutsnuwu*, or "Fortress of the Bears." The ancients regarded the giant animals as supernatural beings. Now, the island is a national wilderness area, thanks to the grass-roots insistence of the modern locals for whom the great bears and their mountainous rain forest also evoke spiritual rather than commercial sensibilities. The old-growth timber on Admiralty would have fallen to the chain saw for export paper products had the people not risen up to save it.

About sixty thousand brown bear, or grizzly, remain in North America, two-thirds of them in Alaska and most of the rest in the Canadian north. The largest concentration of *Ursus arctos,* a variation known as the Shiras grizzly, is found on Admiralty. These bears are typically darker in color and about middle-sized in the range of their species. They are omnivorous, like humans, thriving in habitat rich in salmon, berries, roots, and small mammals.

Juneau, Alaska's capital, is the only town in the country with a drive-up glacier in the city limits. The Mendenhall flows from the vast Juneau Icefield to within a few miles of tidewater, where its melting runoff is carried by a river of the same name. It's on the road system, a standard tour-bus stop, and a natural calendar for locals who watch the receding mass of blue-and-white ice uncover rocks and gravel as its retreat marks the years. This view overlooks a meadow of cottongrass.

The pioneer plants that occupy the scoured terrain left in the glacial wake include these lavender lupin, a remarkably prolific, hardy species that thrives in newly formed, low-grade soil from California to the Arctic. A variety of geranium and chocolate lily share this spring meadow. Like many excesses in Alaska, the spring and summer displays of wildflowers tend to the extreme.

The sense of superabundance and well-being in nature is everywhere in the water paradise. Yellow pond lilies, for instance, are common in the Tongass National Forest, which includes most of Southeast Alaska. The green pods of the water plants are favorites of experienced backcountry travelers who heat them with butter or oil until they swell to become pond lily popcorn.

From April to October, the harbors of Juneau, Ketchikan, Wrangell, and Skagway in Southeast Alaska play host to cruise ships, sometimes dropping as many as five thousand visitors in town on a single afternoon. Until the mid-1970s, only the Alaska State ferries and occasional Canadian ships packed tourists to Southeast Alaska. Since then, a literal boom has transformed tourism in Alaska from adventures in the backcountry to convenient, even luxurious voyaging.

Juneau's history is bound to a geologic extrusion of mineral-bearing rock known as the Gold Belt, a piece of commercial good fortune between the sea and the coastal mountains that triggered a gold rush at the turn of the century. The town's other definitive boom came in the late 1970s when North Slope Crude fueled runaway growth of state government, now the dominant industry in town, with tourism running second.

Most of the rush's high-grade gold production was over by the 1920s, and industrial mining took over at dozens of sites where the remnants of mills, machinery, and labyrinthine tunnels into the hard rock can still be found. A resurrection of gold, copper, molybdenum, and other mining began in the mid-1980s much to the consternation of many locals who favor preservation of wilderness over such environmentally threatening industry.

The Alaska-Juneau Gold Mine shut down in 1944, and the lights of the great low-grade ore mill on Mount Roberts above town never went back on. Most of the equipment was taken from the mine by salvage crews in the 1960s, and the main mill building burned soon after. The honeycomb of mine tunnels remains in the mountain, of course, and the city renovated portions of the mine entrances, including the shops in Last Chance Basin, for a museum.

According to several reports, many of Juneau's young people have fallen in love in cars in the parking lot in front of this waterfall that tumbles down the steep flank of Mount Roberts. Normal life in Juneau includes such handy vistas, midnight picnics on wild beaches, and deserted islands for overnight camps just minutes from town by boat.

The most startling thing about glaciers on first encounter is the cobalt blue ice flashing from the crevices and fractures. Parts of a glacier look blue because the densely compressed ice allows the absorption of all colors but blue in a delightful optical phenomenon. The next characteristic of glaciers that can rattle you a bit is the noise, a crackling and popping and even whistling that suggests anything but ponderous advance or retreat.

Above Muir Inlet, Glacier Bay, Alaska.

GLACIER BAY GUIDES

any years later as he faced the firing squad, Colonel Aureliano Buendia was to remember that distant afternoon when his father took him to discover ice.

—GABRIEL GARCÍA MÁRQUEZ, *One Hundred Years of Solitude*

The memory of moving ice for those who have seen it is indelible. "We were crossing the bay early on a warm evening," one wilderness kayaker told me, "and as we came in front of the glacier we stopped paddling. We were drifting easily with a light wind in the direction of the wrinkled face of it, hypnotized by that wild blue light dancing against the white. I wasn't at all ready for the noise of this thing as big as Los Angeles, cracking and roaring out of the mountains as if it were going fifty miles an hour instead of fifty miles a century or whatever it is. I've never sensed anything before or since that so clearly delivers the message that we are not in control here. I was more than a little bit frightened."

In the summer of 1968, a fairly ordinary man with the unlikely name of Hayden Kaden moved from Texas to Southeast Alaska with his wife, Bonnie, in a Volkswagen Microbus. Four years later, they were leading expeditions to the ice in Glacier Bay, the first such commercial kayak trips into the wilderness that John Muir called "unspeakably pure and sublime." And for the next couple of decades, the Kadens, like environmental evangelists, would purposefully convert their clients to the notion that the earth comes first, that the glory of nature demands stewardship. And they've all had a pretty good time along the way.

As is most often the case with the consequential events of youth, Hayden and Bonnie's migration to Alaska transpired with its beginnings in coincidence, and with no hint of the importance afoot. They had met in France, both on sojourns from college, Bonnie Shoemaker from Ohio State, Hayden from Houston getting college done in Memphis. Back home, Hayden finished his senior year and Bonnie went off to Atlanta to airline stewardess school, but by this time living so far from each other was impossible. In the spring of 1964, after Hayden graduated, they married and moved back to Texas.

The university in Austin accommodated both Hayden in law school and Bonnie in romance languages, and they were near Hayden's mother, Mae, and her second husband, an inspired

Presbyterian preacher named Sunday. Until that point, Alaska was a place encountered in passing through a friend from France who lived in Anchorage; it held no particular allure. Neither Kaden burned with the call of the wild or considered the possibility that they would follow John Muir as ecological proselytizers.

But the dice were rolling in Austin. In 1967, Bonnie had a professor who had worked in Alaska, and that led to the consideration that Hayden might work as a government lawyer in Juneau. As it happened, the fellow running the legislative affairs agency there came to Austin to interview law students for jobs, and so it went.

Nudging fifty in the 1990s, Hayden is a calm, compact man with thick graying hair and mustache, and a kind of background radiation of quiet amusement. Bonnie is his female mirror, fair and blonde, a small, beautiful woman who fairly vibrates with energy most of the time. Both are confident and engaging, delivering a matter-of-fact ease in their interactions with others. They had been attracted to Alaska by the freedom and adventure, but not necessarily by the banquet of the wilderness; remote backcountry was not much of a destination in itself. Nor had they been attracted to Alaska for so-called career opportunities.

"The whole reason for working was to not work," Hayden says now. "We wanted to retire, not to vegetate but to avoid spending our days collecting money doing things we didn't like to do." So they settled in Juneau, then a pre-oil boom town of about ten thousand just a whisper away from territorial days, sprinkled on the shoulders of a fjord in the Alexander Archipelago. Fronted by the sea and backed to the east by rugged mountains and icefields, Juneau eventually demands that a person embrace nature as a condition of tenure. Hayden went to work for the legislative affairs agency as a lawyer, Bonnie for the Department of Education as a publicist.

"It was great," Hayden says. "We met a lot of people we still know, and we bought some land in Gustavus with the idea of moving out there some day. Seven hundred and fifty dollars an acre, no taxes. We worked for two years and then decided to take a year off just to see if we could do it." They roamed Mexico, returned to Juneau in the spring of 1971, deciding to work for two more years and then move permanently to Gustavus in the summer and Mexico in the winter. "We just decided to build a house over there, and neither of us had ever picked up a hammer."

The motivation for so sketchy a proposition as building a house in Gustavus was, in part, the place itself. The village sprouted on a glacial plain, projecting from the rain shadow of the mountains to the north and so enjoying more sunshine than is common to the region and offering a pastoral overture to the staggering ice and peaks of Glacier Bay. The townsite was laid out by a handful of fishermen who had tired of the halibut and salmon grounds nearby and modestly turned to farming in 1914. The rich, flat terrain between Strawberry Point and Point Gustavus looked just right. By 1917, they were producing livestock, rutabagas, strawberries, and other produce that they sold to the canneries around Icy Strait.

Later, during World War II, the construction of a military airfield further stabilized a small population in Gustavus. Excursion Inlet, ten miles to the east, was the site of a major supply depot for the campaign in the Aleutian Islands. The idea was to barge cargo up the Inside Passage to Excursion Inlet, then transfer it to ocean convoys bound for the Aleutians, fifteen hundred miles west. By the time the camp was completed, though, the Japanese and American troops had finished their business on Kiska and Attu, and oddly enough, German prisoners of war were brought to Excursion Inlet to dismantle the camp.

The rest of the commercial history of Gustavus was shaped around the creation of the Glacier Bay National Monument in 1925. The national park lodge at Bartlett Cove is accessible by road from town, which means visitors can jet to Gustavus (since the big military runway is still maintained) for a hiking and sightseeing trip into Glacier Bay aboard a Park Service launch, private boat, kayak, or charter plane. (Any travel in Glacier Bay requires a permit.)

Kayaks are particularly suited for wilderness experiences because they are so inconspicuous; paddlers must submit to wind and tide in a way that no other water traveler does, and as a reward they're almost invisible to the creatures around them. In a kayak, you are among the marine mammals and birds, rather than

on the outside looking in. "The cruise ships are a way for a lot of people to see an area without changing the land, and that's good," Hayden says. "I don't particularly like looking up from my lap when I'm paddling to see one of them, but that's the way it is."

At about the time Hayden and Bonnie decided to build their house in Gustavus, they chanced upon a man named Chuck Horner, a Methodist minister by training who was translating his spiritual vision into wilderness experiences in Alaska. As the 1960s became the 1970s, and we had all seen the photograph of the whole earth taken from an *Apollo* spacecraft, the evolution of ecological consciousness gained momentum. We were shedding the Victorian sensibility that the earth exists just for us, that wilderness can be exploited without limit, restraint, or consequence.

Chuck Horner's *cause célbre* in 1972 was to demonstrate the recreational value of Alaska's parks and fisheries in order to keep logging, mining, and other resource extraction industries in check. Votes are power in a democracy; the more people value wilderness for spiritual comfort and recreation, the less they will be willing to sell it off. Horner approached the Kadens with a plan: He would buy kayaking equipment and teach Hayden and Bonnie backcountry skills; in return, they would become part of his company, Alaska Discovery, and run low-impact kayak/camping trips into Glacier Bay.

"The idea was to save special places, and one way was to take middle-class people to them and point out the beauty and the threat," Hayden recalls. "Before we started on Glacier Bay, we ran trips over on Chichagof and Admiralty islands with Chuck, learning how to work with clients and demonstrating the recreational potential to the Forest Service."

The Kadens bought a surplus army Quonset hut to store the kayaks and other equipment and made their first trip into Glacier Bay in the summer of 1973. By mutual agreement, Bonnie ran the base in Gustavus, booking clients, managing the gear, getting people from the airport to Bartlett Cove, and securing permits. "Bonnie saw the future," Hayden says. "She got the exclusive rights to book commercial kayak trips from the Park Service and that really made everything work. Without the permits, we would have been out of business in short order."

The first trip into Glacier Bay was a comedy. "I was totally green as far as spending time in the wilderness," Hayden says. "My brother Tim Sunday went with me, and the clients were locals from Juneau, I think. We had no rain flies for the tents and it rained the whole time, no wool clothing, duck-down sleeping bags that were soggy and heavy. We paddled in as far as Goose Cove [about twenty-five miles into the outer part of the bay] and never saw a glacier."

People who have traveled with Hayden say his personality is perfect for tending to details and providing the assurances people need when they venture into remote territory. "It took me five years on Glacier Bay to feel as if I knew what I was doing," he says. "After ten years, everything was pretty much second nature.

"The social compact in guiding is complicated," Hayden says. "You are essentially making a living bringing people into a place valued mostly for the absence of people. As far as the clients themselves, you are constantly defining and dealing with the differences in people. Most people are a little bit afraid when they arrive, and you want everybody to have a good time. You want to let people know that they can enjoy themselves in the wilderness, that nature does not have to be serious, that we belong in nature.

"Some of the things you do are relatively straightforward: You provide all the gear so everybody has an equal shot at comfort, chores, and pleasure; you make sure people don't hurt themselves; you try to bring out the good qualities in people. I've run things with one guide for every six or eight clients and added an assistant after that. The guides do all the cooking, planning the paddling or hiking for each day, and make sure we take out everything we bring in. The territory takes care of the rest. It really is rewarding to be with people seeing Glacier Bay for the first time."

John Muir's first glimpse of Glacier Bay on October 24, 1879, came in the company of Hall Young (a Presbyterian missionary) and a crew of Tlingit seamen from the village of Hoonah across Icy Strait. The bay was blanketed by clouds and fog, but two days later, Muir left the party to climb a fifteen-hundred-foot ridge above what is now Geike Inlet. There the sky cleared briefly, and

Muir, astounded, made the following notes:"I saw the berg-filled expanse of the bay, and the feet of the mountains that stand about it, and the imposing fronts of five huge glaciers, the nearest being immediately beneath me. This was my first general view of Glacier Bay, a solitude of ice and snow and newborn rocks, dim, dreary, mysterious."

And ever-changing. As compelling as the spectacular vistas, the wildlife, the sea, and the ice is the revelation that constant change is the primary truth in the landscape of Glacier Bay. What John Muir saw that day no longer exists. Receding ice has doubled the extent of the navigable inlets of the bay; the premier inlet is named after the great naturalist himself. Although we understand the constancy of geologic change and the drama of glacial mechanics in the abstract, we are left in awe when we confront their reality in Glacier Bay. What we see there forever alters the way we consider a parking lot or a suburban shopping mall, a beach or a river.

The people of the ancient villages around Icy Strait confronted the mystery of the ice with myth and struggle. In one revealing instance recounted by Dave Bohn in his book *Glacier Bay: The Land and the Silence*, an elderly chief of the Hoonahs asked Hall Young if he could borrow Young's god for some prayers or if Young would at least pass on a message.

"For what do you wish me to pray," Young asked the man, who was camped with his women at the foot of then-advancing Brady Glacier. "Do you see that ice mountain?" the old man asked, gesturing to the glacier. "Once I had the finest salmon stream on the coast right here, with a deep pool below it to which great schools of king salmon came each spring. To spear them or net them was very easy and they were the fattest and best salmon among all these islands. My household had an abundance of meat for the winter's need, but the cruel spirit of that glacier grew angry with me, I know not why, and drove the ice mountain down toward the sea and spoiled my salmon stream.

"I have done my best. I have prayed to my gods. Last spring I sacrificed two of my slaves, my best slaves, a strong man and his wife, to the spirit of the glacier to make the ice mountain stop. But it comes on, and I want you to pray to your god, the god of the white man, to see if he will make the glacier stop."

Since John Muir and Hall Young made their expeditions into Glacier Bay, thousands of people have submitted to the vision of a world that is still being both created and destroyed, a place that should be preserved for its own sake apart from any justification or rationale. "Part of our whole deal has been to politicize our clients about land use and what is happening to the wilderness, and what to do when they go home," Hayden says. "A lot of them are people who have power, so if I show them how their lives connect to nature, I'm reaching people who can do something about irresponsibility toward the earth. If they got to Glacier Bay in the first place, they're probably caring people, and I just provided a boost."

As the 1980s drew to a close, Hayden Kaden had been guiding around Glacier Bay longer than anybody. He's wintering in Juneau again, lawyering, leading treks in The Himalaya, and had spent some time as a Witness for Peace in Nicaragua with his father, the Reverend Sunday. "My father introduced me to the idea that you can make the world a better place, and you have an obligation to do it."

He and Bonnie have officially been divorced for a few years, but their lives still flow in the same stream. She is the principal of the small school in Gustavus, and their daughter, Sierra, is now a teenager. Alaska Discovery is thriving, offering trips in Glacier Bay, the Arctic National Wildlife Range, Admiralty Island, Hubbard Bay, and wherever people need to be to understand where they are.

I wasn't at all ready for the noise of this thing as big as Los Angeles, cracking and roaring out of the mountains like it was going fifty miles an hour instead of fifty miles a century or whatever it is. I've never sensed anything before or since that so clearly delivers the message that we are not in control here. I was more than a little bit frightened." — Roald Simonson, kayaker, on his first encounter with a glacier. Above is Walker Glacier in Glacier National Park, Alaska.

Kayaks are perfect craft for exploring the wilderness from the sea because they are subject to wind, tides, and currents in ways that make them and their occupants almost invisible. Glacier Bay is a kayaker's dream, with good guides, safe and well-traveled but still pristine routes, and virtually endless side trips ashore.

When Captain James Cook entered Southeast Alaska's Inside Passage in 1778, he sailed past a wall of ice on the northeast shore about forty miles into Icy Strait. When the fog cleared on a morning of exploring for pioneer naturalist John Muir one-hundred years later, the ice had retreated thirty miles inland, revealing, Muir noted, "a solitude of ice and snow and newborn rocks, dim, dreary, mysterious." Today, cruise ships sail sixty miles up into Glacier Bay and kayakers venture even farther.

Grand Pacific Glacier, at the head of Tarr Inlet, is as far as you can go into Glacier Bay in the 1990s, but that's changing every day. Although a few individual glaciers are advancing, the earth's temperature is warming and the icefields of the planet are in general retreat. Glacier Bay National Park, whatever its portions of ice, land, and water, is the largest single unit in the National Park System.

The name *Glacier Bay* orients us toward the Icy Strait entrance to this mystical region, and some imaginative effort is required to grasp the magnitude of the place including its "other" sides. The towering Fairweather Range, for instance, occupies a narrow coastal strip between Glacier Bay and the Pacific in its lower seventy miles and eventually merges with the giants of the Wrangell–Saint Elias Mountains to the north. "Behind" Glacier Bay, too, are scenes like this aerial photograph near the confluence of the Alsek and Tatshenshini rivers.

The cruise liner *Westerdam*, in Johns Hopkins Inlet, Glacier Bay. "I don't much like to look up from my lap when I'm paddling to see one of them," Hayden Kaden says, "but that's the way it is. And the cruise ships are a way for a lot of people to see an area without changing the land, and that's good."

Within Glacier Bay National Park's northern boundaries is the confluence of the Alsek and Tatshenshini Rivers. Both rivers originate in Yukon Territory and flow south through British Columbia and merge to form Dry Bay. It is an area of massive glaciers and steep cliffs as seen here dominated by 15,300 foot Mt. Fairweather.

The Alsek River's looping channels are perfect for spawning salmon, and the annual migration draws wildlife, including fishermen, for the feast. Because the broken rock left by the glacial retreat transforms hard bottom into gravel beds, the salmon can scoop out redds for their eggs and future generations.

On the mainland side of Glacier Bay, the dominant stream is the Chilkat River, a fertile stream with an extended estuarian zone and the kind of tidal inundation that is also just right for salmon and their predators. Several of the many steep tributaries flowing into the Chilkat have their beginnings in the icefields over Glacier Bay.

The moose, the largest mammal on the mainland coast of Southeast Alaska, depends on the flats carved by glaciers and rivers such as those on the eastern side of Lynn Canal. These precious wetlands nourish eelgrass and other fauna for food and also provide the moose and their young with defensible territory.

When I first arrived in Alaska, I recall my inability to look at enough at one time to conclude anything about what I was seeing, so I surrendered in the way I have since learned to relax when listening to symphonic music: Thinking is impossible and a mistake; feeling is everything. My senses steadied from cataloging to meandering, picking up new shades and subtle forms in the overwhelming spectacle.

A view of Haines, Alaska, and the Chilkat Range.

GOLD FOR THE SOUL

Just before dawn on a spring day during my nineteenth year, with no prospects and very little money, I boarded a Juneau-bound ship in Haines. This was the last leg of a continental migration from my Atlantic roots and, like all finales but one, a miraculous beginning as well: Southeast Alaska would be the home I never knew I had until I got there. That morning, I was also certain for the first time that an event of lifelong consequence was at hand; until then, the confusion of my dissipating childhood had so overwhelmed me that I could only react. Reflection had been impossible. As we ghosted from the dock with absolutely no fanfare though, I knew I would remember that moment decades later. And I do.

I recall the feeling of the deckplates shuddering through the soles of my feet as we gained speed, and the way the standing odors of fresh paint and diesel vanished with a chilly snap in the breeze of our motion. I remember looking over the white rail into the water of Lynn Canal, so dark I supposed it was shallow and covering black rock. And then, as though it were yesterday, I remember my first review of the landscape in the faint bloom of morning, which probably accounts for the staying power of my impressions.

As we broke clear of the harbor the light gained the force of distance, revealing a procession of mountains along both sides of the canal as far as I could see. Dark, elementally green forest rushed to the tideline down slopes so steep that I wondered what kind of angle was permissible for such growth. The mountains above were towering, raw extrusions sculpted in shapes of rock, snow, and ice. And because the morning was calm, a duplicate of the dramatic terrain flashed in the still mirror of Lynn Canal, rippling into an impressionist hallucination in our wake.

I recall my inability to look at enough at one time to conclude anything about what I was seeing, so I surrendered in the way I have since learned to relax when listening to symphonic music: Thinking is impossible and a mistake; feeling is everything. My senses steadied from cataloging to meandering, picking up new shades and subtle forms in the overwhelming spectacle. The mountains took on a gray-plum color, and at water's edge a broad band emerged, wrinkled in degrees of brown between the tideline and the high-water mark, underscored by the wild amber shim-

mer of kelp beds in the nearshore wash.

Along many stretches of the tidal juncture I could see that the cobbled beaches disappeared, the forest holding its progress above cliffs of hard, bare rock sewn through with streaks of white quartz. Lone trees occupied surprising precipices like giant bonsai, their seeds apparently nourished by the good luck of all survival to yield an existence clinging to nothing at all. In a few places, streams breaching the mountain facade were marked by alluvial fans that spread broad sandy flats into Lynn Canal.

Overhead, particularly around the streams, soared the first eagles I'd ever seen. Lots of them. On high they caught the full burst of light just coming to me on the surface, emphasizing their bright heads and the white streaks in their feathers. Their sheer size was astounding; eagles are enormous animals to be in the air, some with wings longer than a grown man is tall. But the elegance of their flight was even more astounding. These birds are hunters, carrion eaters if they must be, but preferring, say, a live fish finning on the surface.

As though I had wished it, an eagle stooped in pursuit right when I was watching. From maybe two hundred feet, it tipped up on one wing and sideslipped through most of its altitude, leveling out ten feet above the water. I could see it make minute adjustments to its attitude with individual feathers flickering, using the energy of its plunge rather than flapping wings for this phase of the attack. The eagle flared, extending its yellowish talons into the airstream where feathers on their trailing edges further aided the final stage of the approach.

The eagle hit the silver fish (a herring, although I didn't know that at the time) and, in a motion unlike any other in nature, reached both enormous wings forward, almost touching, and pulled itself from the eruption of water into the air again. Beneath it, the fish struggled on the talons, signaling the end of its life with winks of the sun's reflection. I lost sight of the eagle against the shadows of the lower slopes into which it flew, and I could not believe what I had witnessed. I turned to look for a collaborator to verify the vision, but I happened to be alone on that part of the deck.

Coming on the upper reaches of Lynn Canal for the first time has spun the compass of more lives than mine. This magnificent fjord is the less-celebrated geologic cousin to Glacier Bay fifty miles west and the prelude vista for thousands of seekers who came for the gold. They were refugees from the crumbling Industrial Revolution at the turn of the last century, fleeing the broken promises of a world barely able to feed itself and hopeful that this exotic place could provide the riches to save them. Many settlers, however, stayed in Alaska for reasons of more profound worth.

Lynn Canal is the seaway to the Klondike, a land of instant myth that excited visions of treasure and then fixed them forever with the romantic truths of hardship. The gold stampeders came and found riches, adventure, or pathetic conclusions, but no matter what their ends they received glimpses of lives free of the quiet desperation of most. And if their arrival day was clear, they saw the same place I saw that morning in 1964, and they might have suspected they were up to something other than seeking a fortune.

"Few of us had the inclination to look at the truly grand scenery," wrote Tappan Adney, a journalist from *Harper's Weekly Journal of Civilization*, in 1898, and recorded in William Bronson's *The Last Grand Adventure*. "Snow and glacier capped mountains rising thousands of feet from sparkling water, burying their lofty heads in soft, cottony clouds, are for other eyes than those of miners excited by the preparation for the real commencement of their journey." When they had the inclination, though, as surely they would have on a morning like my first in the territory, they would have known that just seeing Alaska for the first time made them even up to that point; everything from then on was profit.

Lynn Canal is the northernmost reach of the Inside Passage, a fjord etched by successions of glaciers and tidal erosion into the talus slopes of the mountain range we now call the Chilkats. The explorer George Vancouver, a member of the British navy, named Lynn Canal after his own birthplace, Kings Lynn in Norfolk. At its southern extreme, the canal is formed at Hanus Reef where Chatham Strait forks, sending Icy Strait and Glacier Bay to the west and Lynn Canal to the north-northwest. Its channel forks at Seduction Point, and then again before narrowing further and completing its mergers with the four substantial

rivers along its course: the Chilkat, the Chilkoot, the Taiya, and the Skagway. The latter, in the local Tlingit language, means "End of Salt Water."

Lynn Canal's fifty-eight-mile run from Hanus Reef is a navigational horror for the uninitiated or careless mariner. Local commercial fishermen loathe the canal in the fall because of the sudden gales and the dearth of sheltered anchorages. They call it a blow-hole. And the bones of dozens of ships litter the bottom of the deep channel, swept to their fate off shoals like Vanderbilt Reef, Lena Point, and Poundstone Rock. Some carried loads from the gold fields, which still attract the fancy and effort of treasure-hunting divers.

In one celebrated tragedy, the 245-foot British steamer *Princess Sophia* grounded at full speed on Vanderbilt Reef in the winter of 1918. She was on a run from Skagway to Juneau, with an overload of 343 passengers from the gold fields, and behind schedule due to the construction of additional berths to handle the extra people. Boats from Juneau rushed to the scene and rescue of all hands was possible, but the *Sophia*'s captain ordered everyone to remain aboard until another of the company's steamers could arrive in three days to evacuate them. In a gale, before the rescue ship arrived, the *Sophia* broke up and slid off the reef, with the loss of all hands and passengers. The resulting legal tangle lasted fifteen years.

The gold rush attracted miners, entrepreneurs of every stripe, and journalists. Until NASA and space travel, the era was the most heavily recorded in history, save the world wars. Vast collections of photographs remain in albums and boxes, and hundreds of the men and women who went north wrote books of varying degrees of merit. Historians have had a field day with "the madness accompanying gold," noted William Bronson in his book about the gold rush. Between 1897 and 1899, thirty to forty thousand people made their way to the gold fields. Many were the sons and daughters of the Forty-Niners of the previous generation who had dominated the dreamer class of California's settlers. The Civil War was fresh in their memories, and the Spanish-American War with all the bluster of Teddy Roosevelt was under way. And, despite William Jennings Bryan's eloquent speechmaking, he was losing the presidential election to William McKinley.

One of the big issues was Bryan's support of the proposal to take America off the gold standard, known as "bimetalism" in those days. Centralization of capital, foreign demands on U.S. gold reserves, and a demoralized labor force were crippling the country's manufacturing engines. Hundreds of thousands of Americans were homeless in 1896, and farms were failing because the farmers couldn't get enough cash for their crops to keep up with their mortgages. "You shall not press down upon the brow of labor this crown of thorns," Bryan said at the Democratic convention that nominated him for his losing candidacy. "You shall not crucify mankind upon a cross of gold."

To a lot of people, the obvious answer to the gold crisis was to produce more gold, and no greater impulse could have existed for a man foraging for survival on the clams of Puget Sound than the prospect of becoming a millionaire. Alaska's earliest strikes were around Juneau in 1880, but those rich placers quickly gave way to what would be, until 1946, the largest hard-rock gold mining complex in the world. The mines of the Juneau Gold Belt were low-grade ore, and tons had to be wrestled from the mountains, crushed, and sluiced to produce a single ounce of gold.

In 1898, as the Juneau strikes were moderating into steady industrial production with hard labor more available than bonanzas, the Klondike hit. The stampede began when the steamship *Portland* docked in Seattle carrying the legendary two tons of gold from the Klondike, an event still celebrated all over the Northwest and Alaska. To grasp a sliver of this golden hope, a stampeder could choose from two primary routes to the interior of the Alaska-Yukon wilderness.

The long all-water alternative was by ship across the Gulf of Alaska, through Unimak Pass in the Aleutians, into the Bering Sea, and up the Yukon River from its mouth at Saint Michael. The voyage could take from three months to several years. Later, the black sand beaches of Nome and the upper Bering Sea region proved bonanzas as well. The other, more accessible route took stampeders up the Inside Passage (the name originated during the gold rush), past Vancouver Island, through the Alexander Archipelago of Southeast Alaska, and finally, up Lynn Canal.

There, the gold seekers unloaded their belongings on the river beaches and prepared for the terrible overland journey to the Klondike. A grim, predatory society arose in that climate of urgency and greed as the ships disgorged thousands of passengers on the tidal flats of Dyea or the long-wharves of Skagway. At Dyea, the beginning of the Chilkoot Trail, stampeders had to race the twenty-five-foot bore tides with their belongings and then face the merciless Chilkoot before reaching the navigable, though still difficult, plateau at the top of the pass.

Only twenty-seven miles as the crow flies, the terrain was so steep it took days or weeks to cross. Then, the miners had to pack from four hundred to eight hundred miles to the gold fields. One of the most notorious sections of the Chilkoot Trail, known as "The Scales," produced a world-famous photograph of an unbroken line of men trudging up the 30-degree slope on the fifteen hundred "Golden Steps" carved in the ice and snow. Some editors even increased the angle when they ran the picture so the men appeared to be struggling up an almost vertical wall. Sizzle sells papers.

From Skagway, the miners submitted to White Pass, which later, when the stampede became industry, was the route of the Yukon-White Pass railway from the gold fields to tidewater on Lynn Canal. Other routes during the stampede included terrible mistakes overland from Edmonton and through Valdez off Prince William Sound, but they produced little but suffering and death.

News of the stampede drew dreamers from around the world who in most cases arrived too late to cash in on the real money, so they became laborers in an industry that thrived in Alaska and the Yukon well into the twentieth century. The placer mines of the interior resisted heavy industrial methods for the most part, though huge camps and platoons of workers sprouted from Dawson City to the giant dredge operations near Fairbanks and Nome.

In Southeast Alaska, the gold trade peaked and flourished for an entire generation in the 1920s and 1930s when about thirty major hard-rock mines were going full blast from Windham Bay forty miles south of Juneau on Stephens Passage to Berners Bay midway along Lynn Canal. Here, the miners threw power at the gold, building massive stamp and ball-crusher mills to work the ore into the gravel and powder that could be sluiced. A society of managers, miners, and suppliers grew out of the wilderness and lasted until 1946 when the fixed price of gold couldn't keep up with the high cost of digging it out of the mountains.

I arrived in Juneau on the night of the day that began this reminiscence. A month later, as it happened, I was working as a laborer on the crew dismantling a gold mine, the last operating vestige of the gold rush. The Alaska-Juneau Gold Mining Company still maintained an office in town and had just sold a salvage company the rights to pull the thousands of miles of copper wiring out of the mine, along with the hoists, rails, and other heavy equipment. I made more money than I'd ever thought possible, went to college, did the things we all do as our lives form, but I never left Southeast Alaska for too long. And it wasn't the money that kept me coming back.

Skagway, in the Tlingit language of Southeast Alaska's ancient settlers, means "End of Salt Water," and for the stampeders of 1898, it meant the beginning of a terrible overland passage to the Yukon gold fields. Many prospectors were refugees from a collapsing national economy down south, who hoped to make a stake; many were simply determined to live a life apart from the ordinary.

Modern Skagway's celebration of the gold rush is their prime stock-in-trade, and the entire town has been refurbished inside and out over the past twenty years. The bars, hotels, and shops sport turn-of-the-century decor, and you can watch resident actors reenact the Shooting of Soapy Smith, the outlaw lord of the town for a few years during the boom.

The wild years of the gold rush calmed in two decades, and an industrial approach to wresting precious metal from the rocks took over, including a miracle of a railroad from the gold fields to tidewater. The narrow-gauge Yukon–White Pass railway carries only tourists now, but once it was a lifeline for miners and a route to market from the high-volume mills that bloomed in the interior.

Between 1897 and 1899, thirty to forty thousand people made their way to the gold fields. Nobody was counting. The Civil War was fresh in their minds, the Spanish American War with all the bluster of Teddy Roosevelt was under way, and a nationalistic fervor gripped the nation. These days, the midsummer party on the Fourth of July is still an annual highlight in the small towns of Southeast Alaska.

Preservation of the past in Southeast Alaska has not been limited to the recent incursion of the European boomers and their followers. The northern flats and estuaries of Lynn Canal were extremely rich in fish, game, and edible fauna, so the Tlingit bands of that region were particularly powerful, with highly evolved forms of expression. At Haines now, resident teachers offer instruction in the ancient arts and crafts of totem and silver carving, dancing, storytelling, sewing, and basketry at a native cultural center.

The Chilkat Mountains, near Haines over Lynn Canal. Coming upon the upper reaches of Lynn Canal for the first time has spun the compass of more lives than mine, that's for sure. This magnificent fjord is the less-celebrated geologic cousin to Glacier Bay, fifty miles west, and the prelude vista for thousands of seekers who came for the gold. In the winter the road through a valley in the Chilkats from Haines Junction to Haines is the only road into Southeast Alaska from the outside. A summer road gets you to Skagway from British Columbia at Carcross, but that's it. Haines is the northern terminus for the Marine Highway ferries and, for a lot of people who arrive overland, the first and often most lasting vision of Alaska. The town itself owes its beginnings to a Presbyterian missionary, Hall Young, and his traveling companion, John Muir.

Few of us had the inclination to look at the truly grand scenery," wrote journalist Tappan Adney in 1898. "Snow and glacier capped mountains rising thousands of feet from sparkling water, burying their lofty heads in soft, cottony clouds, are for other eyes than those of miners excited by the preparation for the real commencement of their journey."

The gold stampeders came and found riches, adventure, or pathetic conclusions, but no matter what their ends they received at least glimpses of lives free of the quiet desperation of most. And if their day of arrival was clear, they saw the same place I saw the morning I arrived in 1964, and perhaps suspected they were up to something other than seeking a fortune.

Mount Drum, in the Wrangell-Saint Elias National Park, Alaska.

CAPE SPENCER TO COOK INLET

Seward Harbor, on the east coast of the Kenai Peninsula, Alaska.

FAIRWEATHER TROLLER

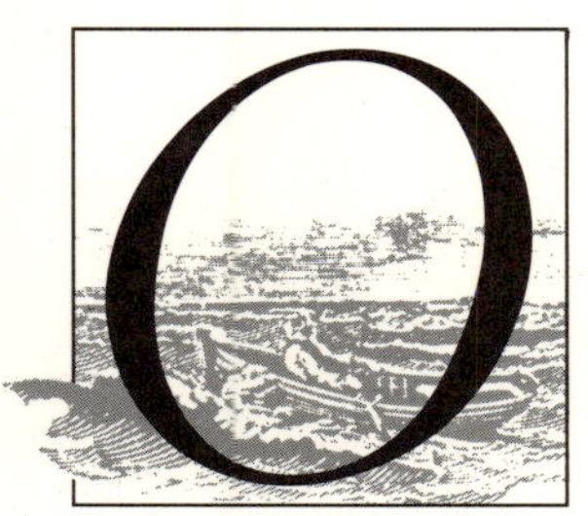

On harbor days in the 1920s when the sea was too rough to fish, the salmon trollers hunkered in their boats in the timbered coves of Southeast Alaska, whiling away the hours making lures. They snipped rough shapes of spoons from sheets of brass and copper, placed the metal over hollows carved in ironbark blocks and, with small ball peen hammers, tapped in subtle curves until hunch and experience told them to stop. The spoons imitated baitfish — the herring, candlefish, and caplin that attract king salmon, the trollers' main prey.

Those early lures were crude; no two were alike, but they caught fish and freed the trollers from the more tedious work of always fishing with bait. And during lulls in a storm, the old trollers say, the *tink*, *tink*, *tink* of hammers hung in the air over the anchorages, mingling with the odors and fragrances of the rain forest. A few descendants in the commercial trolling fleet cherish a handmade spoon or two, although most of the designs are mass produced in wide varieties.

Practiced from California to Southeast Alaska, trolling is an esoteric form of commercial salmon fishing; the fish are taken individually by hooks and lines set in intricate patterns from slowly moving boats. Both the boats and the fishermen are called "trollers." Troll-caught salmon are prized by smokers and lox makers because they are killed immediately, bled and cleaned, and then carefully iced or frozen while at sea. No better salmon reaches the table.

Many trollers fish alone, true ocean hunters with highly evolved maritime skills and a profound understanding of the animals they hunt. Until the mid-1970s, the trollers roamed the Pacific coast year-round and delivered to classic trolling ports like Fort Bragg, California; Neah Bay, Washington; and Craig, Elfin Cove, Sitka, and Pelican, Alaska. Now, the grim equation of too many fishermen chasing too few fish has forced limitations on the seasons to conserve the salmon stocks, so the troll seasons are weeks instead of months long.

One of the most seductive aspects of trolling, especially before the current era of regulations and limits, was exploring for new grounds. Trollers are secretive, only sharing their discoveries with a few close friends, called a "code group" because they talk in code on their radios. "Fish are where you find them" is an old saw

among trollers, which means they can be anywhere. The dream, of course, is to find them where nobody else has.

Toivo Anderson, a Finn, ended up trolling in Southeast Alaska after a Bering Sea ice storm drove his father's immigrant family from its first settlement in Nome, two thousand miles to the north. Toivo was seventeen years old when he took to the trolling grounds in 1927, fishing first from a boat he named *Flower*, using cotton lines, bait, the hammered spoons, and the genetic savvy of a bred mariner. He fished his way to a new boat, the *Yakima*, whose previous owner, Fred Wooleson, known as *Yakima* Fred, was only slightly less of a legend than Toivo himself. A troller is often known by a combination of his boat's name and his given name, like *Yakima* Fred, *Chasina* Chip, and *Caribou* Barry.

In 1947, Toivo built a brand-new boat, the *Greta*, and aboard her three years later made a discovery now considered the most memorable among the hook-and-line fleet. Running with his Finnish pals, Toivo pioneered the Fairweather grounds, a patch of the Pacific over a pair of undersea mountains between twenty-five and fifty miles off the coast dominated by Mount Fairweather. The sea mounts rise from the depths, stimulating a great upwelling of nourishment that extends from plankton, through small fish like herring, to salmon, and then to the sea lions and other mammals that prey on them.

Though other trollers had done well closer to shore along the Fairweather Coast, no one took the chances involved in running eight to twelve hours offshore into the Gulf of Alaska, arguably the meanest piece of ocean on earth. The weather blows up on the gulf in minutes, and there is no place to hide. The only protection is Lituya Bay, in the middle of the Fairweather Coast, a pocket harbor with a wicked bar across the entrance that has driven more than one fisherman to seek different work. In July 1958, Lituya Bay was the site of a massive tidal wave that denuded the surrounding mountain slopes to an elevation of 1,720 feet.

To fish the Fairweather grounds, you run out, fish, turn off the engine and drift while you sleep, load your boat, and, if you're lucky, make it back to Pelican to deliver. Likely as not, you'll blow out a wheelhouse window, lose your rigging, or miss the fish completely and return to port empty. Despite the risks, Toivo and his pals joined the lunch line on the Fairweather grounds and for years had it all to themselves. Toivo remembers that on his best day he landed 282 large king salmon, all over ten pounds and most over twenty. A good day for a troller anyplace else is 50 kings. When the word leaked out, as it always does, about the huge loads of king salmon coming from "somewhere out west," others followed.

Troller and author Frank Caldwell celebrates the Fairweather Coast in *Land of the Ocean Mists*. "This is a soul-searching, mind-boggling chunk of real estate, this Fairweather country," Caldwell writes. "No wonder the Tlingit natives loved it, so much so that for centuries they coexisted with its great glaciers bulldozing, mountainsides toppling, avalanches hurtling, and ice groaning."

Everything about the Fairweather Coast urges caution to an approaching visitor, though all who draw near forever alter their notions of what transpires when the continent meets the sea. Geographically, the region is a 150-mile stretch between Cape Spencer at the northern end of the Inside Passage, northwest to Ocean Cape, which guards Yakutat Bay. Geologically, the zone is defined by the Fairweather Fault, where the North Pacific and North American tectonic plates collide in an aria of mountains. This land reminds us that the earth, although made, is still being made, as naturalist John Muir once said of Alaska.

At a rate of 1,000 vertical feet per mile, the terrain rises from talus beaches to the peak of Mount Fairweather at 15,300 feet and is broken by other spectacular peaks; Mount La Perouse, Mount Quincy Adams, Mount Root, and Mount Wilber top 10,000 feet, as do a half dozen more, all within twenty-five miles of the sea. On the inland side, the parade of geologic giants falls away into Glacier Bay, whose waters come within thirty miles of the Pacific where John Hopkins and Gieke inlets pinch into the Fairweather Range.

The few inlets along the coast are familiar to fishermen as scenes of big scores of salmon or of blessed protection from the brutal Gulf of Alaska. From Cape Spencer, the *Coast Pilot* names Graves Harbor, Torch Bay, Dixon Harbor, Astrolabe Bay, and Palma Bay within twenty-five miles, and then nothing, no inter-

ruptions in the wall of mountains, until Lituya Bay. Northwest from Lituya, there is no shelter for seventy miles, past Cape Fairweather, its massive glacier flirting with the Pacific at tidewater, until Dry Bay and the mouth of the Alsek River.

The opening of the Fairweather grounds was like filling in the already wondrous life of the trollers with background music by a symphony orchestra. Catching salmon with hooks continues to seduce people along the Pacific Coast, and most agree that making a living by tricking great game fish like king and silver salmon into biting has an allure beyond the price of the fish. Trollers describe their work in terms of common sense, mystery, enthusiasm, art, and luck but always with caveats of humility and modesty. Ask any troller, whether a legendary highliner like Toivo Anderson or a novice, what kind of spoon, bait, plug, or technique works best, and he'll almost surely begin his reply with, "Well, I don't really know much about it, but . . ."

What a troller does know about goes something like this: It is dawn on the Fairweather Coast (remember the orchestra) aboard a wooden boat that is small by ocean standards, maybe forty feet, powered by a steady six-cylinder diesel. Like many trollers, this one fishes alone on his boat, although some carry deckhands known as boat pullers. The run through the gray light of arctic night is just another wheelhouse interlude with coffee and a cigarette, followed by a resolution to quit, and then breakfast off the oil stove, which also nudges the chill out the sliding door and onto the deck. Visibility is a few hundred feet in dense, low-lying fog; yesterday, the fog burned off early to release a bright, sunny morning.

Eating, the troller watches his radar screen and loran receiver — electronic tools that Toivo Anderson would have given anything to own. The loran receives signals from powerful transmitters on shore, triangulates them automatically, and tells the troller where he is according to special charts. He has been drifting, asleep, for the four hours of his short night, and now the loran is ticking its way back toward yesterday morning's numbers where he caught twelve large kings before eight o'clock.

Finally, the numbers tell him he is where he will set his lines. He finishes his coffee, stands up, grabs a handful of hard candy from a can, and reaches under his chair to throw a handle that shunts hydraulic fluid through hoses to the small winches used to haul the lines. Then he leaves the wheelhouse.

The deck is slightly awash, the ocean soughing through the scuppers with the roll of the boat. Trolling boats ride low and wet, because the fisherman must reach each fish as he brings it alongside. The boats are also distinctive with their forty-foot-long poles extending from amidships; a big spring attaches four or six main wire lines to the poles, and a tin bell fastened to the spring signals the sometimes tentative strike of a salmon.

The troller works from a cockpit in the stern, which places his midsection right at the waterline, giving him the leverage he needs to pull the big fish aboard. From the pit, he surveys the sea at first light, seeing birds nearby — shearwaters and white-belly petrels — a good sign. His senses quicken as he takes his place in the food web as a hunter. His lures and baits are carefully lined up on the shelf behind the pit. (With few exceptions, trollers are compulsively neat about their gear.)

The baits — herring, because he thinks nothing is better in the early spring — were readied the night before and stowed just forward of the pit in a blue, ten-gallon trash can like the one Sesame Street's Oscar the Grouch lives in. The troller's daughter gave it to him for good luck, and it has lasted three seasons already. And from the rack over the pit, he takes his morning spoons, two of the old, heavy Superiors he won in a card game from Radio Eddie's father, who'd had them for thirty years. The troller never uses them when he thinks there are sharks around, and he believes they bring him good luck.

He sets his lines as the sun breaks over the Fairweathers, brightening the surface fog and dancing on the sea along the wind rips and the rippling wake. His boat moves at two knots, riding a gentle westerly swell. He pops a hard candy into his mouth, settles his hips on the back of the pit, and notices for the first time how chilly and damp his oilskins feel.

Waiting for his first bite, he talks to himself, "Okay, we're fishing now. Yes. Good morning." And he looks over the deck and forward to the wheelhouse to see that everything is neat and set up just as he likes it. He steers with a tiller in the pit. "Come on salmon, bite the hook," he chants, another part of his ritual. He'd heard a sport guide do that when he was a kid, just before he'd first felt the astonishing power of a king salmon almost tear a fishing rod from his hands. The troller looks around at the birds; they're hitting the water right where he is, materializing through gaps in the fog that is already burning away.

The white crags of the Fairweathers are breaking free, the peaks emerging first as the fog hangs over the beaches on the horizon toward shore. "Come on, salmon . . . " And then the fish is there. First, the tin bell signals tentatively, *tink, tink*, and then the whole spruce pole is rattling. The spring is stretched way back. He can feel it along the bulwark actually shaking the boat, and he can see the wire from the pole slashing the water. This fish is so big it lifts the fifty-pound cannonball fixed to the end of the wire where the monofilament lines stream with their lures and baits.

His other lines are taking hits, too, but nothing like the big one. He begins to haul his starboard lines and finds that the fish is on the bottom leader—a good sign that this is a giant. Staring into the black sea, he waits for the salmon, and then it is there, as big as a shiny fifty-five-gallon drum. "Yes," he says, "yes." He has taken hundreds of big kings over forty pounds, but this one has him cranked up a little.

The salmon is on a fifty-foot-long leader, still strong and easily cruising faster than the boat. The brass flash of the spoon in its mouth is clearly visible; then the fish senses the boat and dives. A thick rubber link between the leader and the wire stretches out, quivering, but it looks as if the fish is well hooked. "Sharp hooks, always sharp hooks," the troller thinks, not even aware if he spoke out loud, his mind now totally with this big salmon. He waits for the fish to tire against the resistance of the boat and line and the rubber snubber, and ten agonizing minutes pass when the salmon can pull free at any time. The troller must land the fish by pulling hand over hand on the monofilament leader until he can reach the salmon with the gaff.

He tries to bring the fish near, but it is as though the line is eyebolted to a wall; he lets the salmon soak some more, waiting and nursing the cuts in his hands that won't heal all season. Finally, he gains line against the fish, which is suddenly on the surface and alongside, just ten feet away, still cruising abeam. But now the fish is bleeding, and it comes easier, rolling on its side to flash white and gold in the morning light that seems to be a part of it rather than reflected from it. The troller anxiously surveys the water near the boat, hoping not to see sharks or sea lions, either predator would gladly fight him for this big fish. There is nothing threatening.

Then the salmon is alongside, and the troller can smell him; king salmon have an odor unlike any other creature's, a deep, musty smell suggesting weeds and earth. The troller reaches down familiarly for his gaff, hanging in the pit. "Easy now. Easy." He leans as far as he can toward the fish and, with a practiced touch, raises the great salmon's head slightly, snapping the heavy end of the gaff down on a spot about an inch back of the big wild eyes. And the salmon dies. With the last of the same motion, the troller sticks the metal point of the gaff into the fish's head and hoists it aboard and into the bin ahead of the pit. He has to use both arms and then can only barely lift the huge fish.

The salmon is beautiful, maybe seventy pounds in its four feet of length and two feet of breadth. Its head and back are deep black, dissolving like shades of lacquer from ebony through a brightening spectrum of dense tones to the silver of its broad flanks. Its scales are the size of dimes, its gill plates like sheets of polished steel.

The troller reaches into the bin and slaps the meaty side of the fish, grinning before he realizes how odd such pleasure seems alone at sea. Then he turns the salmon, takes his spoon from its jaw, and gets to work on the other lines. The salmon will be worth three hundred dollars at 1989 prices, but the money is far from the troller's mind as he turns aft to catch the flash of bright sun on the snowy flanks of the Fairweather Coast.

CAPE SPENCER TO COOK INLET

The Kennecott copper strike at the turn of the century in the Wrangell–Saint Elias Mountains was nearly the equal of the gold rush in value. The scenery in the territory was terrific, for those who made it, including not only the meandering spectacle of the Copper River but a breathtaking range of massive peaks that rise from tidewater to touch the sky. Many are over fifteen thousand feet high.

McCarthy, Alaska, a village of just a dozen or so people now a few miles from Kennecott itself, was one of the centers of the rush to the Copper River bonanza with a bustling population of miners, families, and attendant characters. The Copper River boom created another town, Copper City, now called Cordova. It was the tidewater port for steamships hauling goods in and copper and gold out.

Cordova today is a fishing town, home to most of the fleet that works the Prince William Sound and Copper River Delta salmon grounds. This slough at Point Whitshed is typical of the salmon habitat that once made the sound so prolific. Now, a substantial portion of the fishermen's annual catch comes from fish hatcheries they built to defeat the cyclical fluctuations that in certain years almost broke them.

The arc of coastline of the magnificent inland sea we call Prince William Sound joins the wild Fairweather Coast of the Panhandle of Alaska to the interior, shedding the rain forest along the way. Inland, across the Chugach Mountains on the northwest edge of the sound, the terrain, flora, and fauna are distinctly subarctic rather than maritime.

This view of a peak in the Chugach Mountains is from Thompson Pass. The actual highway from Prince William Sound to the interior of Alaska now runs from Valdez, through Thompson Pass, skirts Mount Wrangell at the northern end of that chain, and drops onto the plateau that in two hundred rolling miles descends to Cook Inlet.

The caribou, an unlikely critter with its out-of-balance rack of antlers, ranges across the tundra flats throughout the interior, beginning north of Prince William Sound. The herds migrate on north–south routes over hundreds of miles, and the vision of thousands moving through a pass, steadily for days, can change forever the way we perceive land mammals.

When the Trans-Alaska Pipeline became part of our cultural idiom in the early 1970s, one of the biggest environmental hurdles was providing passage for the caribou and other migrants along the route. The ecological disruption of the pipeline was further emphasized in 1989 when the tanker *Exxon Valdez* was run aground on Bligh Reef in Prince William Sound, spilling 11 million gallons of pipeline crude.

"Fragile" barely begins to describe the relationships in the maritime and subarctic ecosystems that are dominant in the range of this book. Short, intense seasons of growth, life, and death yield the specific flora and fauna necessary to the overall functioning of our bioregion, and no amount of puzzling will give us answers to questions about how all the pieces connect, or what happens when we lose one like old-growth fir or estuarian grasses. Without a doubt, prudence dictates extreme conservation, especially on the Northwest Coast.

When people finally stand up to oppose something like destroying wilderness and say 'enough is enough,' it is not always a logical matter. It is mostly spiritual," Han Timmers told me. "I think everybody wants to live in harmony with their place when they get time to think about it."

Max McCarty and Noel Pallas, Prince William Sound fishermen, with a pair of Copper River salmon, photo by Brad Matsen.

FLAT CRAZY

Aboard the salmon gillnetter *Christine Sue* on the Copper River Flats, the jazz violinist on the cabin tape player fights a losing battle with the spring gale. The Gulf of Alaska is delivering a full ration, a sixty-knot screamer, hammering the boats hunkered in the meager shelter of a sandbar known as Egg Island. The fishermen, Max McCarty and his mate Noel Pallas, set their net despite the storm, and after a few minutes on deck looking for the reassuring splashes of snared salmon, withdraw to the relative comfort inside. They will let their net soak for a couple of hours as the arctic night turns into its only slightly grayer cousin, the day.

Visibility outside is a quarter-mile or so in driving rain and spray; you can see just barely farther than the end of the gillnet suspended from an undulating line of white corks. The nine-hundred-foot net is strung between a pair of bright orange buoys, fluorescent anomalies against the silty brown water of the river delta. A good haul for a single set on the flats is one-hundred fish; a great set is four hundred; legends are made of seven-hundred-fish sets, which can literally sink a boat.

The season opens and closes during May in hours-long bursts, so most of the fleet denies the weather and fishes. Each fish is worth about $20.00, an astonishing $2.60 per pound. At that price to the fisherman, Copper River reds and kings are selling in supermarkets for $9.00 to $11.00 per pound, and on restaurant plates for $10.00 to $30.00 per six-ounce serving. But the king and red salmon that mob the three-hundred-thousand-acre delta of the Copper River are the Beaujolais Nouveau of the annual spawning migration that give the coasts of the Pacific Rim their abundant character.

They are the first and, many say, the best of the new crop. Their rich color and high oil content quite simply define "salmon" for discriminating chefs, gourmet diners, and backyard barbecuers. And though all salmon have high levels of artery-clearing omega-3 oils, these from the 430-mile Copper River have among the highest levels. The longer the river, the higher the oil content since the fish need the nourishment their bodies carry to battle upstream to their ceremonies of birth and death. Startling longevity among the aboriginal bands of the delta is a matter of local myth, and one man told me some ancients lived to two hundred years because of their diets.

Copper River salmon harmonize careful handling by fishermen and processors, ingenious marketing, and the miracle of the jet airplane. A diner in a four-star restaurant in Seattle, New York, or Los Angeles, for instance, can savor a meal of Copper River salmon caught seventy-two hours earlier on the flats. Such a meal rivals the banquets of ancient Roman nobles, which featured exotic and never-before-tasted fare that was carried by teams of runners from newly conquered territories.

Max and Noel and the other gillnetters of the Copper River Flats begin an elegant chain of events. The world inside the cabin of the *Christine Sue* contracts into a 10-by-12-by-7-foot box, since the wheelhouse windows are fogged by vapor from the working coffeepot. Gourmet coffee beans are sold now in Cordova, *Christine Sue*'s remote Prince William Sound home port, and the rich aroma of Max's blend mingles with the more pedestrian odors of cigarette smoke, diesel fuel, and damp working clothes.

In the aft port corner of the cabin, drying work gloves and hats are clothespinned to twine strung over the oil stove, and they sway against the roll as the *Christine Sue* jogs in the wind-driven chop. Perched behind the steering wheel on a bench he crafted in his woodshop, Max reaches forward to the cluttered dashboard under the windshield and fumbles for an ashtray. Under the working rubble typical of all fishing boats — tools, paperbacks, tidebooks, a *People* magazine with a picture of Donna Rice on the cover, candy bars, knives, a tin of bag balm for sore hands — he finds the ashtray and stabs out his cigarette.

"Gillnetting," he begins, shooting a look across the cabin at Noel, who is sitting over coffee with his elbows on the galley dinette, "is the sport of teenagers." They crack up, laughing with the high tones of an inside joke. "Yeah," Noel presses on, "it used to be that the last bastion of the liberal arts major was law school. Now it's gillnetting." More laughs.

Max and Noel, both of whom have chewed through the better part of four decades, are fishing together this season on the *Christine Sue* because Noel's boat burned and sank last year. Noel works as a bartender in a chain hotel in Elyria, Ohio, in the winter, and has been fishing out of Cordova in the summers since the early 1970s. Max lives in town year-round, something of a rarity among the three thousand or so gillnetters who make annual migrations to Cordova. They come from around the world in a remarkable blend of styles, some still languishing in the blue-collar traditions of European fishing cultures, others of the American New Age, turned out in full-Patagonia and filled with bulletproof optimism.

The town echoes the diversity of its migrants. Situated on the sheltered shore of Orca Inlet, hard against the tidewater edge of the Heney Mountain Range, Cordova is a visual feast. Rare clear days, when the cobalt blue sky draws the mountains even higher from the silver-black sea, have been known to summon lifelong commitments from people. Later, when the rains beat day after gray day on the town, they wonder just what the hell they're doing in Cordova.

Cordova began life as a railroad town, connecting the rest of the world with the Kennecott copper mines, via steamship at the deepwater port. Originally called Copper City, the town was incorporated as Cordova in 1909. By the time the mines closed in 1938, fishing had taken hold as the region's staple industry and so kept Cordova from joining the ranks of ghost towns. About three thousand people live in Cordova year-round now, although the salmon season just about doubles the population. Almost everybody came from someplace else.

Max McCarty, for instance, is originally from eastern Washington, went to Whitman College, then eased through the counterculture days in a farmhouse in Walla Walla. He is married to Heather, a writer, and together they have borne Hannah and Miranda, their daughters. "I've had as much fun as anybody," he says. "I graduated from college with an economics degree. I cooked with a woodstove. I drove a pickup truck with a big engine. I'll tell you, it just doesn't get any better than this," he laughs, waving his arm in an arc to indicate the cabin of the *Christine Sue*.

"I had just always wanted to go fishing in Cordova," Max goes on. "I had a friend at Whitman who came up here in the summer so I came with him and got into a seining operation. I couldn't believe it. Then I made ninety-one hundred dollars in three weeks in 1976 and bought into the boat as a partner.

"Anyway, I moved my family here in '78 so when bad years came I did whatever I had to to keep things going. I worked on the dam at the hatchery over at Port San Juan (the largest salmon hatchery in the world). I'm not a natural fisherman or hunter-gatherer. I'm a natural builder, but fishing is the same in that you get paid for how hard you work. That's okay. People try to live normal lives against the background of the flats," he says. "But at best it's schizophrenia."

When the *Exxon Valdez* changed Cordova, Prince William Sound, and the world forever on March 24, 1989, Max and legions of fishermen from Cordova fought to save their piece of heaven. Although most of the stinking mass of crude inundated the west side, a hundred or more miles from Cordova, the fishermen rushed to protect the hatcheries they had spent so long building. And they did it; they kept the oil out and the salmon came back.

On this morning behind Egg Island though, we talk and rest while the net soaks, drifting and dreaming as if in the lyric of a fisherman's folk song. The *Christine Sue* shudders as the wind delivers a particularly violent blast; a big wave smacks the hull with a power that suggests it has risen up out of the Pacific with the sole purpose of finding us. The sea transforms relaxation to fright in an instant, even though we are "inside" and somewhat protected.

Egg Island is one of several barely emergent barrier bars created over centuries by the silt and sand washed from the river. The Copper carries more sediment than any river in Alaska, even more than the Yukon, and it drains more area into the Gulf of Alaska than any other watercourse. This is largely due to the glaciers that make up vast portions of the Copper's shoreline, grinding and transporting the land to the sea in the timeless waltz of the ice.

However, one spring morning in 1964, during a great earthquake, the topography of the flats was altered dramatically in ten minutes when the floor of the delta rose six feet. The once-familiar channels, passages, bars, and fishing holes simply disappeared, and new ones took their places. Fishermen had to learn to navigate over the treacherous sands all over again to reach the Racetrack, Poulsons, Walhalla, Gus Stevens, Pete Dahl, Softuk, Kokinhenik, and the other well-known drifts on the Copper River grounds.

The Copper River, so named because of the massive copper lode that brought the mines to the valley, flows from headwaters on the talus slopes of the eight-thousand-foot Mentasta Mountains, 430 miles northeast of the delta. From there, the river first carves an uncertain meander through a soggy high plateau and tails into a narrow glacial assertion through the Wrangell-Saint Elias range of the Chugach Mountains.

On the seaward slope of that spectacular geologic statement, the Copper becomes a torrent. All rivers eventually assert themselves in an alluvial fan at their mouths, but some, owing to the composition of the land through which they make their final plunge, or to the power generated by the steep gradient, fall to the sea in a brutal, scouring rush. So it is with the Copper.

On July 20, 1741, just south of the barrier islands, an event transpired that changed the human politics of the coast forever. The ship *St. Peter*, commanded by Vitus Bering, anchored off a low strip of land we now call Kayak Island, took on fresh water during a stay of just twenty-five hours, and sailed back west. Bering, a Dane in the service of Russia, would die with most of his crew in the Komandorskiye Islands on the way home, but his record of the voyage launched a fur rush that eventually became the colonial outburst we remember as Russian America.

The *St. Peter*'s doctor, naturalist Georg Steller (who survived), had gone ashore with a boat crew and recorded the existence of human beings as evidenced by still-warm cooking fires. He had also ascertained — incorrectly as it turns out — that the large blue jay on Kayak Island was the same species as a bird found in the Carolina colonies on the Atlantic. His conclusion at the time, however, was that the North American continent was a whole piece accessible from the Pacific. The record of the voyage further contained the first evidence in white European annals of the existence of Prince William Sound, its barrier islands — Hinchinbrook and Montague — and the great Copper River Flats.

Fishing the flats is a matter of understanding gross tidal hydraulics and simply staying alive. The flood tide brings the fish over

the shallows, but they disperse over three hundred thousand acres; the ebb leaves them stranded and moving in the channels that surround the higher terrain, and there the fishermen set their nets. The best fishing — and the most danger — is found where the breakers pound the sandbars. Once, when Max was fishing in his skiff, he was trapped in the surf by a deadly current, which threatened to dash the boat to pieces. To save himself, he built a fire right on deck hoping another boat would see him and know that he was doomed if he wasn't rescued. It worked, and one of his brother fishermen got a line to him.

The flats have been declared "nonnavigable water" by the Coast Guard, which therefore takes no responsibility for marker buoys. The depths and locations of the bars change every year though, so fishermen get together to mark the channels themselves. The markers are poles cut from straight trees and planted on the channel edges; radar reflectors are hung from the poles for help in the fog, and a map of the new channel courses is available each spring from the fishermen's union office.

Many, many who fish the flats have seen the toss go the wrong way, and some die in the sea every year. Even the careful ones. The day before this May opening, a couple of dozen drivers climbed into their turbine-powered race cars at the Indianapolis 500, statistically a far safer proposition than fishing Copper River salmon. "When you're fishing the flats," the saying goes, "you're either bored to death or scared to death. But business is business."

As the morning brightens, the flats reveal themselves as long shadows running in gray horizontal layers. The wind and shallow sea seem to be of the same substance, rather than separate from the sands that take the eye as far as it is able to see. Picking up the net begins an hour before the one-day fishing period is to close. On deck, Max wears a full-body suit of orange nylon, Noel is in logger's heavy trousers called "tin-pants" and a fleece pullover.

From the brown water comes the net, wound by Max and Noel on a reel with the help of burping hydraulic power, the corks thumping over a roller mounted on the stern of the *Christine Sue*. The reds, weighing about six pounds each, come over easily in the net; the kings, though, some of which weigh thirty or forty pounds, have to be eased over by hand. Noel stands by with a gaff to catch those that drop from the net back into the water.

The total for the set is not good, about fifty reds and ten kings. But just fifty-six hours from the moment the fish are brought aboard the *Christine Sue*, they will be served to restaurant customers all over the country. For quality's sake, the salmon are bled immediately, carefully iced in the hold, and delivered to the fishermen's co-op in town. At the co-op plant, the salmon are gilled, gutted, packed in ice, and taken to the Cordova airport for the four-hour jet flight to Seattle.

From the Seattle airport, a seafood forwarder picks up the salmon and delivers them to a route of restaurant and supermarket customers, many of whom buy full-page ads in local papers to announce and celebrate the arrival of the Copper River salmon. The arrival of these fish from a remote river on the American Pacific has become a rite of spring, a symbol of well-being and achievement, and firm proof that the biological furnace is operating as it should. "I don't think I could do anything else," Max tells me.

The Copper River is the vein of the Wrangell–Saint Elias ranges, carrying the water of rainfall and melting snowpack to the sea, where it begins its nourishing return. The salmon return upriver to offer themselves as food for birds, mammals, and lower organisms that participate in soil production. Even the ponds of accumulated water in the Copper's drainage form from rain and runoff over the high water table of rock and permafrost and breed the mosquitoes to feed the birds that sing, amaze, and even feed us.

The possibility that a sense of delicate balance can emerge after an hour's observation of a rocky cliff doesn't sink in quickly. Seeing a tree or sprig of green clinging to a cliff gives us a hint that we really don't know how all this works, even if it's possible to find a botanist to empirically sort through the mechanics of such adhesions. The main message of wilderness observation always seems to be that humans are not on top of anything, least of all the webs of nature, and we must be careful.

The Kenai National Wildlife Refuge, where this vision of a pond captured photographer Pat O'Hara, is 2 million acres of mountains, tundra, and two large lakes, within easy reach of Anchorage. Popular for kayaking, camping, and wandering, the refuge is also home to the largest population of moose anywhere.

This is Horsetail Falls, near Valdez, home of the Trans-Alaska Pipeline terminal, a monstrous installation that literally lights up the side of an entire mountain at night. Alaskans are fond of saying that a town like Valdez or Anchorage or Juneau is just fine because it's only a few minutes away from Alaska.

The peaks of the Chugach Mountains are the "local" mountains for Anchorage and Cook Inlet, separating the subarctic flats from the Gulf of Alaska at Prince William Sound. Keystone Canyon, near Valdez, is typical of the extreme variation in vegetation over relatively small altitude changes that characterize the Chugach because of its northern latitude.

North of Prince William Sound, Alaska becomes the arctic equivalent of a desert, with far too little precipitation at sufficient altitude to create glaciers. It takes the ocean's moisture to do that, and even the icefalls of Denali can't compare to the glaciers flowing out of the Chugach, Wrangell, and Saint Elias ranges to tidewater, like the massive Columbia shown here.

This view of the Kings Inlet area of Prince William Sound from the Kenai Mountains looks pristine after a fresh snow fall. However, during the summer of 1989, the volcanoes on the southwest coast of Cook Inlet errupted dropping ash on the region. Only the highest peaks and those on the eastern slopes were spared the smudge of ash. In the long run the ash is beneficial, providing nutrients for plant growth.

The Kenai Fjords National Park, on the south end of the peninsula of the same name, was the subject of much concern in the spring of 1989 when the stinking mass of crude oil from Exxon's tanker invaded that coast. The park is dominated by the Harding Ice Field, which discharges tidewater glaciers into the Gulf of Alaska and is responsible for the sculpted fjords and inlets that equal those of Glacier Bay.

"I suppose the best thing you can say about ice other than that it chills our drinks is that it freezes time," said an insightful cynic about water's solid form. Seasonally, breakup in Alaska reveals a forgotten autumn's debris, along with the miracles of lakes and streams reborn; in its long-running incarnation as a glacier, retreating ice gives us a vision of our home planet hundreds and even thousands of years ago.

When Mount Redoubt erupted in 1990, Anchorage, two hundred miles north, went dark in the afternoon. The chain of active volcanoes that begins across Cook Inlet from the Kenai Peninsula continues for eight hundred miles across the Aleutian Islands, and then for another several thousand miles down into the Japan archipelago. Part of the sense of wilderness in Alaska is the magnitude of scale, especially after you leave the more comforting confines of Southeast Alaska and Prince William Sound.

This is Portage Lake at the head of Cook Inlet between Anchorage and Seward on the Kenai Peninsula. At this point a railroad from Seward and Wittier ties the sea to the 450,000 square miles of mainland interior Alaska, and tentatively asserts the human presence into one of the earth's last great wildernesses. Occasionally, a storm on Turnagain Arm or a wicked freeze shuts down the railway, but that's all for the better. It wouldn't do to get too comfortable with the idea that we're running things here.

Columbia Glacier, from Prince William Sound, Alaska.

EGG TAKE ON A TIN RIVER

At the head of Sawmill Bay, steep slopes of fractured rock and the excavations for the hatchery give the otherwise pristine cove the raw, scarred look of a small strip mine. As on most such islands that define the ragged western edge of Prince William Sound, the cobble of the beach gives way to tentative stands of spruce and hemlock typical of the northwest extremes of the Pacific forest. Sawmill Bay is on Evans Island, one of a couple of dozen mountainous fragments that guard the mainland coast and part of a series of remote inlets known collectively as Port San Juan.

The hatchery on Sawmill Bay was conceived and built by the salmon fishermen of Prince William Sound who were unwilling to submit to the natural cycles of good years and bad years, with their unrelenting lack of concern for human fortunes. By the mid-1970s, the methods for large-scale hatchery production had been perfected, so the fishermen set to work altering nature. They taxed themselves to raise money, saw to it that laws protecting their rights to catch most of the returning fish were passed, and picked a site for their tin river.

"We had nothing we could count on," says Jerry Thorne, who has fished the sound for fifty years. "We could starve any year before the hatcheries, but the people here did it. I started fishing in a skiff in 1938 when I was eight years old. My father fished in a sailboat on Bristol Bay in the Bering Sea, and he told me to stay here, on Prince William Sound. 'It's a death trap on Bristol Bay,' he told me. 'Stay here.' He was right, but without the hatcheries, we'd be nowhere."

To hatch in the spring, fertilized salmon eggs must be held through the winter in conditions that mimic those of a real river; therefore, the most critical commodity at a hatchery is pure, fresh water. Larson Creek, at the head of Sawmill Bay, has not hosted a natural salmon run since it was blocked a few hundred feet up the beach by a high cliff, which the salmon cannot pass — the result of glacial and geologic indifference tens of centuries ago. From this creek, though, the salmon growers could draw water for a hatchery and, barring a drought, sustain the eggs for the winter.

As the salmon leave the hatchery-rearing pens in the spring, they are marked by the sense triggers of their spawning stream — in this case, the aluminum raceways — and so return at the end of their ocean grazing to spawn on a tin river. Until they reach the

tin river, though, they are fair game for fishermen who catch all but the small percentage required to support the hatchery by providing both eggs for the next generation and cash to meet expenses. This system is known as salmon ranching because the fish take to the open range, as compared with salmon farming, where the fish are reared from fry to harvest in captivity, thus providing no benefits to the fishermen. Salmon farming is illegal in Alaska.

The hatchery at Port San Juan reached international celebrity status in March 1989, after the tanker *Exxon Valdez* altered nature in a different way by spilling its cargo of North Slope Crude into Prince William Sound. The tanker grounded on Bligh Reef, sixty miles to the northeast, but tides and currents quickly carried the stinking mass of oil right to Port San Juan, threatening the hatchery just days before the year's fry were to be released into the sea.

And of the hundreds of millions of fry that had been released the year before, 46 million or so were due to arrive in two months for harvest or spawning, and nobody really knew what the oil would do to them or where the salmon would encounter the drifting crude. If you throw a bottle into Prince William Sound off Port San Juan and it doesn't go aground anywhere, it will probably drift in an enormous circle on the North Pacific, passing Kodiak Island, the Aleutian Chain, then around to Sitka and Yakutat, and back to the sound. This supercurrent, known as the Pacific gyre, is the heart of the salmon's nourishment web, the equivalent of a giant pasture.

Within hours of the spill, fishermen were racing to protect Port San Juan and their other hatcheries on the west side of the sound. They deployed makeshift booms and used their boats to skim the oil from the water in a desperate attempt that eventually succeeded, making the fishermen their own heroes. "It's all over if the oil gets into the pens," one told me. He was racked from lack of sleep after days fighting to save Port San Juan. "People don't realize what it takes to get something like a big hatchery going. We'll have to start all over, and this time we'd know that this can happen so there'd be a lot less to get excited about."

When I first witnessed an egg-take at Port San Juan in 1981, the hatchery was in its second year and about to realize its potential as the most productive in the world, so there was plenty of excitement. The pink and chum salmon fertilized and hatched that summer and released to the sea the following spring would equal the natural run of pink salmon to the sound on their return a year later. The fishermen had vowed they would never again remain tied to the docks because harsh winters, unfavorable conditions at sea, or other vagaries of nature claimed the salmon before they did.

At the first big egg-take, the heady sense of control over nature was palpable at Port San Juan as the fishermen made good their vow. Even the crews working sixteen-hour days taking the eggs and milt by hand were captured by the power of such a dream come true. When the hatchery corporation got around to designing souvenir shirts, the most popular depicted a leaping, smiling salmon and the motto: "We hatch 'em, you catch 'em."

Just the sight of the salmon returning in force for the first time accounted for much of the enthusiasm, as it has every summer since. Hundreds of thousands of the five-pound silver packages of protein rolled in fertile splendor in the holding pool below the aluminum raceways — tin rivers they had to mount in the genetic dance of spawning. It didn't matter to them that the gravel redds of a stream and death's natural embrace had been replaced by aluminum and the efficient human ceremonies conducted by crews brought to the hatchery for two weeks of taking and fertilizing the eggs.

Spawning with humans as midwives is a frantic, hopeful, delicate, and exhausting business, fraught with uncertainty and the odd sense that nature is grimacing at our mechanical, self-serving urgency. For two weeks, the tedium of long days of repetitive work produces scenes like this: Along a series of wooden boardwalks, ramps, and stairways from the incubator rooms to the tin river, a man in a plaid shirt zigzags down the hill carrying a pair of white five-gallon buckets, and disappears into a big plywood shed at the water's edge. He emerges minutes later and runs back up the hill, now taking care not to spill the water and red eggs that show clearly in the plastic buckets, backlit by the afternoon sun.

The lanky egg carrier seems to have invented a spidery lope to cushion the shock of his footsteps, using his whole body in the same way you'd use your hands as gimbals to steady a cup of coffee on a boat in heavy seas. He moves quickly back up the hill, but the sound of his footfalls on the planking is lost against the wild roar of a hard rain hitting the buildings, the blast of a waterfall over the cliff a hundred yards away, and the flumes feeding the raceways. The only sounds breaking through the concerto of water come from the plywood shed: sharp, unfamiliar smacking noises, occasional shouting, and Percy Sledge singing "When a Man Needs a Woman" on a tape player cranked up so loud that most of the words are lost in the shuddering bass.

Many in the crew performing the numbing work of taking the eggs fill in the lost lyrics of the famous song, enjoying the irony of a barroom love song at an egg-take. The salmon make their way upstream in the raceways to a holding trough in front of the shed; there, workers called killers take them and dispatch each with a single whack of a club to the head. Then, workers called either spawners or buckers depending on whether they perform their duty on females or males, take the stuff of life from the salmon, required in a ratio of three females to one male.

"We have surplus bucks so we're more careful with the females. That's the way it is," a woman bucker tells me, laughing at the sexual politics of the situation. The eggs and milt are squeezed from the salmon into the white buckets, which are then rushed up the hill by the man in the plaid shirt. Once they are mixed in the buckets in the shed, fertilization has taken place — creation, if you like — and those eggs must be at rest in the incubating tanks within five minutes. The sperm, contained in the milt, which is a mixture of several vital fluids, dies within a minute of leaving the fish, but by then it has penetrated the eggs in the bucket.

As soon as the fertilized eggs are in the water of the incubators, they swell in volume by 20 percent in just two hours. Next comes a tender stage, which lasts until the eye of the embryonic salmon appears in six weeks; then the embryo is ready to feed itself on the self-contained nourishment of its yolk sac. By midwinter, the eyed eggs look less like the larval form of insects and more like fish, and as darkness and chill winds press upon Prince William Sound, a new generation of salmon has assumed ancient forms, despite the artificial beginnings.

Long after the killers, spawners, buckers, and the man with the bucket leave, a small force of skilled aquaculturists remain at the hatchery, monitoring and controlling the habitat in the incubators to conform to the rigorous demands of nature. "I would say it turns me on, making all this new life," one told me of her work. "And of course I get to live here."

Here is Prince William Sound, a spectacular statement of the life-force of earth, of birth, death, and renewal nowhere made more clearly than on its western edge where the Pacific washes through barrier islands and into the mainland. Port San Juan is one of thousands of interruptions in a coastline backed by the glacier-clad mountains of the Kenai Peninsula, making the region a paradise for adventurous mariners and kayakers. In recent years, on cruises from the ports of Seward and Whittier, even the somewhat less hardy have had the opportunity to sample what that hatchery biologist and the wilderness travelers experience. No matter how you get there, though, the territory is inspiring.

Carved by ancient ice that has retreated over the past ten thousand years, deep marine intrusions extend into the peninsula from its root in the general vicinity of Anchorage. Southwest from Cape Resurrection on the Gulf of Alaska, the ragged coast continues its magnificent chorale of mountains, ice, and water to the Chugach Islands and Kennedy Entrance, the southern beginning of Cook Inlet, which forms the western shore of the Kenai. Much of this zone has been designated a national treasure and set aside as a wilderness area.

The Kenai Peninsula is formed by a single dominant mountain range running along a southwest-by-northeast axis, merging with the Chugach Mountains to the east of Anchorage, Alaska's metropolis. Looking west from the Kenai across Cook Inlet, you can see active volcanoes, the violent ornaments of the Aleutian Range, an extension of the backbone of the Great Land, the Alaska Range. West of those mountains, the subcontinent of Alaska becomes more modest hills and flats, through which meandering rivers carry out their part of the water-to-water cycle of life.

Like those mountains and flats on the larger scale, the mountains of the Kenai both dominate and create the landmass of the peninsula. From the air — say, on a flight from Anchorage to Homer, 250 miles south on Kachemak Bay — the pattern of ice, water, and drainage is obvious. On a clear day, the mechanical process of flowing water shaping land reveals itself in the loops and braids of runoff streams, which in some cases become rivers and lakes, carrying the accumulations of water in the ice and snow of the mountains to the sea and depositing silt and soil on its way. Thanks to the scouring and accretion of land on the western side of the peninsula, towns like Kenai, Soldotna, and Homer are possible.

On the steep, still-glaciated eastern slopes, though, the Kenai Range loses its ability to translate mountains into earth, and so human settlement has been temporary or abbreviated. The most distinct gap of lower elevation in the range terminates on the sound in the town of Whittier, for years the premier deepwater port leading to the interior of Alaska. During the Great Pacific

War, Whittier was the tidewater terminal for a railway leading to the Alaskan front in the defense of North America, and in the rush to protect themselves at a time when no cost was too great, Americans built a tunnel through the highest portion of this portage, which still connects Whittier to the interior by rail as far north as Fairbanks. Now, the route is put to more benign use as a gateway to Prince William Sound for kayakers and sightseers, bringing them a hundred miles closer to the archipelagos and western coast than the other break in the ice and mountains that allows the existence of the town of Seward. The western sound is not a place for novice wilderness travelers since aid or rescue is not close at hand, even though the powerful urge to view the place pushes many to tempt fate. And in recent years, anyone with the price of admission can board a sightseeing boat at Seward or Whittier and tour the coast and the Kenai Fjords, which explains in part why the tragedy of the *Exxon Valdez* hit home. All who sample the wondrous beauty of Prince William Sound revel in its mere existence, thankful that such places are not substantially altered by human presence. Pristine wilderness is evidence that allows us the solace of a higher power, something greater than ourselves, whether we happen to actually be in a place like Prince William Sound or are just thinking about it.

When I witnessed the egg-take at Port San Juan, I thought, "We're changing the natural order here by creating salmon. What does all this do to natural selection with its requirements of strong years and weak, to the long-range possibilities for survival of the species?" And I had similar thoughts, but with a far heavier heart, as I watched fishermen drag absorbent booms around in Port San Juan trying to save the hatchery from the drifting crude that was also altering nature in 1989. Salmon hatcheries are built and celebrated because they provide food for a hungry world; likewise, shipping oil around allows us to transport those salmon and to drive to the supermarket to buy them. Both are evidence of our voracious appetites, and our willingness to tilt the natural order to satisfy our desire to live as only nobles lived in past eras.

The difference between raising salmon and spilling crude oil, though, arises from their effects on our collective self-esteem: Rearing salmon to feed ourselves clearly is tinkering with the biospnere, but it settles upon us as a creative gesture, whereas spilling oil being shipped to feed our wasteful energy habits upsets our vision of ourselves as responsible stewards of the earth and calls to question our powerful position in the natural order.

If any single human characteristic can be considered essential, it is our ability to make choices that extend beyond instinct, that support the spiritual conclusions we must reach to view our lives as positive. And neither salmon runs from tin rivers nor oil spills destroy the earth; our home planet has demonstrated its resiliency by evolving through far more exotic and catastrophic disturbances. We are left only to decide whether tin rivers and oil spills destroy our visions of ourselves.

BIBLIOGRAPHY

BOOKS

Allen, John Eliot, Marjorie Burns, and Sam C. Sargent. *Cataclysms on the Columbia.* Portland: Timber Press, 1986.

Alt, David D., and Donald W. Hyndman. *Roadside Geology of Washington.* Missoula, Mont.: Mountain Press, 1984.

Bohn, Dave. *Glacier Bay: The Land and the Silence.* San Bruno, Calif.: Alaska National Parks and Monuments Association, 1967.

Brier, Howard M. *Sawdust Empire.* New York: Knopf, 1958.

Bronson, William. *The Last Grand Adventure.* New York: McGraw-Hill, 1977.

Browning, Robert J. *Fisheries of the North Pacific,* rev. ed. Anchorage: Alaska Northwest, 1974, 1980.

Caldwell, Francis E. *Land of the Ocean Mists.* Edmonds, Wash.: Alaska Northwest, 1986.

Clark, Ernest D., and Ray W. Clough. *The Salmon Canning Industry.* Seattle: Blyth & Co., n.d. (Date uncertain, probably about 1930.)

Doig, Ivan. *The Sea Runners.* New York: Atheneum, 1982.

———. *Winter Brothers.* New York: Harcourt Brace Jovanovich, 1980.

DuFresne, Jim. *Glacier Bay National Park: A Backcountry Guide to the Glaciers and Beyond.* Seattle: The Mountaineers, 1987.

Eiseley, Loren. *The Immense Journey.* New York: Vintage, 1946.

Eppenbach, Sarah. *Alaska's Southeast: Touring the Inside Passage.* Seattle: Pacific Search Press, 1983.

Fraser, Hermia Harris. *The Arrow-maker's Daughter and Other Haida Chants.* Toronto: Ryerson Press, 1957.

Freeburn, Laurence, ed. *The Silver Years of the Alaska Canned Salmon Industry.* Anchorage: Alaska Northwest, 1976.

Geraghty, James J., et al. *Water Atlas of the United States,* 3rd ed. Port Washington, N.Y.: Water Information Center, 1973.

Gross, M. Grant. *Oceanography.* Englewood Cliffs, N.J.: Prentice-Hall, 1972.

Haig–Brown, Roderick L. *The Living Land.* New York: Morrow, 1961.

———. *Measure of the Year.* New York: Morrow, 1950.

———. *River Never Sleeps.* New York: Morrow, 1946.

———. *Timber.* New York: Grosset & Dunlap, 1942.

———. *Return to the River.* New York: Morrow, 1941.

Kazmann, Raphael G. *Modern Hydrology,* 2nd ed. New York: Harper & Row, 1972.

Leopold, Luna B., and Kenneth S. Davis. *Water.* New York: Time, Inc., 1966.

LeWarne, Charles Pierce. *Utopias on Puget Sound 1885–1915.* Seattle: University of Washington, 1975.

Lung, Edward B. *Black Sand and Gold.* New York: Vantage Press, 1956.

Márquez, Gabriel García. *One Hundred Years of Solitude.* New York: Harper & Row, 1970.

Michener, James A. *Alaska,* New York: Random House, Inc., 1988.

Morgan, Murray. *The Northwest Corner.* New York: Viking, 1962.

———. *The Last Wilderness.* New York: Viking, 1956.

———. *The Columbia.* Seattle: Superior Publ., 1949.

Muir, John. *Travels in Alaska.* New York: Houghton Mifflin, 1915.

Netboy, Anthony. *Salmon: The World's Most Harassed Fish.* Tulsa: Winchester Press, 1980.

Reisner, Marc. *Cadillac Desert.* New York: Viking, 1986.

Satterfield, Archie. *Chilkoot Pass, Then and Now.* Anchorage: Alaska Northwest, 1973.

Service, Robert. *Collected Poems.* New York: Dodd, Mead, 1966.

Spude, Robert L. S. *Skagway, District of Alaska 1884–1912, Building the Gateway to the Klondike* (occasional paper). Fairbanks: University of Alaska, 1983.

Stewart, Hilary. *Indian Fishing, Early Methods on the Northwest Coast.* Seattle: University of Washington, 1977.

Stone, David, and Brenda Stone. *Hard Rock Gold . . .* Juneau: Juneau Centennial Committee, 1980.

Sundborg, George. *Hail Columbia.* New York: Macmillan, 1954.

Walton, William C. *The World of Water.* New York: Taplinger, 1970.

PERIODICALS

Alaska Geographic. The Alaska Geographic Society, Anchorage, Alaska.Titles include *Alaska's Glaciers; British Columbia's Coast; Glacier Bay, Icy Wilderness; South/Southeast Alaska.*

National Fisherman magazine. Journal Publications, Rockland, Maine.

Northwest Coast was
produced in association with the publisher by
McQuiston & Partners in Del Mar, California:
art direction, Don McQuiston; design Don McQuiston
and Joyce Sweet; mechanical production,
Joyce Sweet and Kristi Paulson Mendola;
copyediting, Robin Witkin;
composition, TypeLink; text type, Goudy Old Style;
printed in Hong Kong by
Dai Nippon Printing Co., Ltd.